MOSSES

AND OTHER BRYOPHYTES
AN ILLUSTRATED GLOSSARY

BILL AND NANCY MALCOLM — MICRO-OPTICS PRESS

First published 2000

ISBN 0-473-06730-7

Design, layout, and illustrations by Bill and Nancy Malcolm.
Printed by Brebner Print, Napier, New Zealand.
Published by Micro-Optics Press, a division of Micro-Optics Ltd.,
Box 320, Nelson, New Zealand

abbreviations and symbols used

• — illustration (on the same page unless noted otherwise).
xs — cross-section.
ls — longi-section (vertical section).
wm — whole-mount.
UV — ultraviolet.
note: the colors of some of the microscope views are "false", produced by re-fluoresced ultraviolet light or by polarizing and differential interference optics.

to Barbara Polly

acknowledgments

Friends, family, and colleagues kindly supplied literature, commented on the manuscript, and helped to collect and determine specimens, and we gratefully acknowledge their help—Sabrina Malcolm, Phil Garnock-Jones, Susan, Kevin, and Claire Archer, Mal and Chris Weston, Jessica Beever, Patrick Brownsey, David Glenny, Rodney Lewington, Darea Sherratt, John Braggins, Victoria Hollowell and two anonymous reviewers at Missouri Botanical Garden, Reijel Gardiner, Paula Warren, Allan Fife, Dale Johnson, and especially Barbara Polly, to whom we affectionately dedicate the book.

The main literature sources we consulted in compiling this glossary were:

Crum, HA; Anderson, LE (1981): *Mosses of Eastern North America*. Columbia University Press, New York.

Flowers, S (edited by A Holmgren) (1973): *Mosses: Utah and the West*. Brigham Young University, Provo.

Hébant, C (1977): *The Conducting Tissues of Bryophytes*. J. Cramer, Vaduz.

Magill, RE (editor) (1990): *Glossarium Polyglottum Bryologiae*. Missouri Botanical Garden, St. Louis.

Niklas, KJ (1997): *The Evolutionary Biology of Plants*. University of Chicago Press, Chicago.

Raven, PH; Evert, RF; Eichhorn, SE (1998): *Biology of Plants* (Sixth Edition). W.H. Freeman, San Francisco.

Schofield, WB (1985): *Introduction to Bryology*. Macmillan, New York.

Schuster, RM (editor) (1983-4): *New Manual of Bryology*, Volumes 1-2. Hattori Botanical Laboratory, Nichinan.

Scott, GAM; Stone, EG; Rosser, C (1976): *The Mosses of Southern Australia*. Academic Press, London.

Smith, AJE; Smith, R (1978): *The Moss Flora of Britain & Ireland*. Cambridge University Press, Cambridge.

introduction

Mosses and other bryophytes were first studied centuries ago when nobody understood plant life cycles and when communication among researchers was haphazard. As a result, the meaning of some of the many terms and phrases that have been coined or borrowed over the years to describe bryophytes are confusing, overlapping, or even contradictory. The original meanings of some have changed, too, either as a particular "school" came to dominate bryology or as our scientific understanding improved, and even today the meanings of some terms are hotly disputed. For such contentious terms, we've tried to include any alternative meanings that are still widely used (they're given numbers in parentheses). In addition, for some terms we've listed others which have the same, similar, or opposite meaning (they're inside parentheses after explanatory words and phrases such as *synonym, compare with, similar to, not the same as*, and *the opposite of*). We've cross-referenced the glossary by printing in **boldface** most of the terms used in the definitions that are defined separately elsewhere. We've also provided more than one illustration for terms that describe highly variable structures such as the spines and papillae of leaves, or brood bodies and leaf cell lumina. Those extra illustrations also contribute toward our other goal of celebrating the remarkable diversity of bryophytes. As well, we've tried to include several illustrations of bryophytes that often are studied in the laboratory of botany courses—for example the moss genera *Bryum*, *Funaria*, *Polytrichum*, *Sphagnum*, and *Tortula*, and the liverwort genera *Chiloscyphus*, *Marchantia*, *Metzgeria*, and *Plagiochila*.

The glossary isn't meant to be a text. Although we tried to make it comprehensive, we couldn't include in the definitions all the inevitable exceptions and other nuances, such as the myriad types of protonemata, apical cells, and branching patterns. Anybody wanting more detailed information should consult the literature listed in **further reading** on page 213.

We've alphabetized the glossary entries by letter only, without regard to hyphens or blank spaces, so for example the entry **lateral** is followed by **lateral-intercalary** and in turn by **lateral leaves**.

The glossary covers mosses, liverworts, and hornworts. We've chosen those three groups because even though they're not closely related, their structure and life cycles are similar enough that a term which applies to one often applies to the other two as well. We've left out the lichens entirely, because although they can resemble mosses and thus be mistaken for bryophytes, they're really fungi living intimately with algae and/or cyanobacteria, and so in our view the terms describing them belong in a separate glossary.

MOSSES AND OTHER BRYOPHYTES AN ILLUSTRATED GLOSSARY

Lopidium concinnum **abaxial** surface of frond. 1 cm

a- (or **an-**) — a prefix meaning *not* or *without* (Greek, *synonym* in Latin **e-** or **ex-**).

ab- (or **-abs**) — a prefix meaning *from*, *away from*, or *out of* (Latin, *synonym* in Greek **apo-**).

abaxial — facing away from a stem or **axis**, hence a leaf's "underside", "outer", "back", or "lower" surface (•left) (*the opposite of* **adaxial**).

abortive — defective or imperfect development.

abrupt — changing suddenly, as in leaves sharply **constricted** to a narrow **acumen**.

acaulescent — lacking a stem (•below, middle) (*similar to* **sessile**) (*the opposite of* **caulescent**).

accessory chromosome — an extra **chromosome** or a chromosome fragment resulting from abnormal **meiosis** or **mitosis** (*synonym* **B-chromosome**).

accrescent — increasing in size from the base to the tip (•below, left).

-aceous — a suffix meaning *belonging to the family*..., for example jungermanniaceous.

achlorophyllose — said of a cell or structure that doesn't contain any chlorophyll (*synonym* **echlorophyllose**) (*the opposite of* **chlorophyllose**, containing chlorophyll) (•below, right).

Bryum billardierei **accrescent** leaf. 0.5 mm

Sphagnum sp. immature **acaulescent** capsules on pseudopodia. 0.5 mm

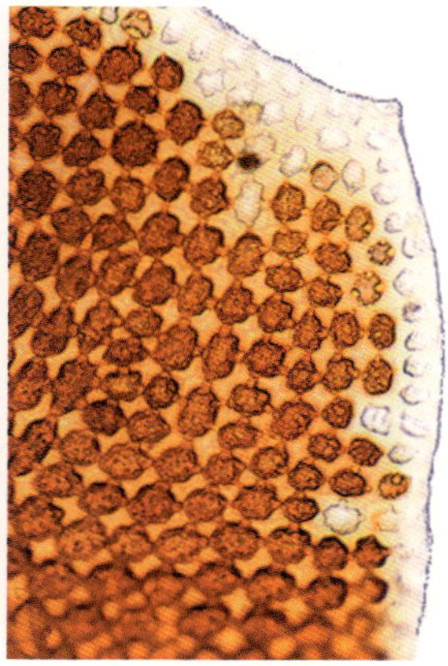

Jamesoniella colorata **achlorophyllose** margin. 50 µm

Acrocladium chlamydophyllum **acostate** leaf. 100 µm

Polytrichadelphus magellanicus **actinomorphic** shoot. 5 mm

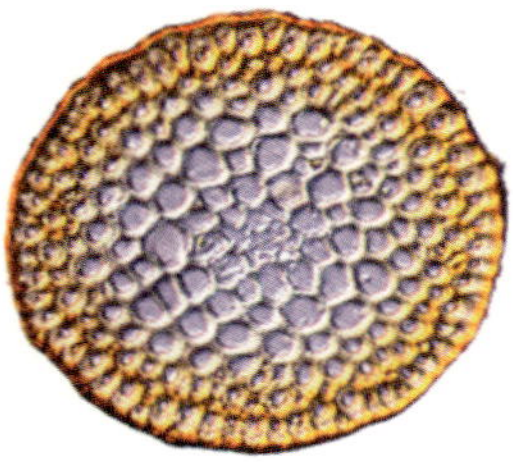

Rhaphidorrhynchium amoenum **actinomorphic** stem (xs). 50 µm

achrophilous — preferring an **alpine habitat**.

acicular — needle-shaped.

aciculate — irregularly scratched as if by a needle.

acidic — said of a **substratum** reacting as an acid or a "hydrogen-donor" and with a pH of less than 7.

acidophilic — preferring an acid **habitat** (*synonym* **oxylophilic**) (*the opposite of* **basiphilic**).

acinaciform — having the form of a scimitar, an oriental sword with a curved blade that broadens toward the tip.

acostate (or **ecostate**) — in mosses, lacking a **nerve** (•left, top) (*synonyms* **enervate** and **nullinervate**) (*the opposite of* **costate**).

acrandrous — producing male sex organs (**antheridia**) at the stem apex.

acrocarpous — producing female sex organs (**archegonia**) and later the **sporophytes** at the tips of the main stems (*the opposite of* **pleurocarpous**). Most acrocarpous mosses are **erect**, **tufted**, and sparingly branched, while most pleurocarpous mosses are **prostrate**, **matted**, and highly branched.

acrodendrophilous — thriving in tree-tops.

acrogyny (adjective **acrogynous**) — the production of **archegonia** from the **apical** cell of a **shoot**, resulting in the loss of that cell and therefore a halt to the shoot's further **vegetative** growth (*compare with* **anacrogynous**, producing archegonia laterally on a stem or branch from superficial initials and without the loss of apical cells). Most leafy liverworts are acrogynous, whereas **thallose** liverworts typically are anacrogynous.

***Acromastigum*-type branching** — in leafy liverworts, a type of branching in which the branch replaces half of an **underleaf** (*compare with* ***Frullania*-type branching**, in which the branch replaces the **postical** half of a *leaf* (rather than an underleaf), ***Microlepidozia*-type branching**, in which a branch replaces the **antical** half of a leaf, and ***Zoopsis*-type branching**, in which the branch replaces the entire leaf).

acropetal — (1) referring to an ascending series, (2) said of tissues or organs developing or maturing from the base towards the **apex** or from the center towards the **margin**, with the youngest tissue at the apex or margin (*the opposite of* **basipetal**).

acroscopic — (1) facing the **apex**, (2) referring to the triangular surface of a **merophyte** in direct contact with the **apical** cell (*the opposite of* **basiscopic**).

actinomorphic — **radially symmetric**, said of a structure that can be cut into two mirror halves along more than one plane (•left, middle and bottom) (*the opposite of* **zygomorphic** or **bilaterally symmetric**). Radial symmetry is considered more primitive than bilateral symmetry.

action spectrum — the spectrum of light wavelengths which trigger some response or drive an energy-requiring process such as photosynthesis. Shade-loving bryophytes use red-skewed light more efficiently than do bryophytes adapted to full sun.

aculeate — covered with prickles (•below, left).

acuminate (noun **acumen**, plural **acumina**) — tapering to a long narrow point (1) gradually (•below, middle) or (2) **abruptly** (•below, right).

acuminose — tapering gradually to a flat, narrow tip (*compare with* **acuminate**, tapering to a long narrow point).

Chiloscyphus aculeatus **aculeate** leaves. 1 mm

Taxiphyllum aomoriense **acuminate** leaf apex. 100 µm

Chandonanthus squarrosus **acuminate** leaf apex. 100 µm

acute — tapering with more-or-less straight **margins** to a point with an angle less than 90° (•below).

ad- — prefix meaning *to*, *toward*, *facing*, *near*, or *at*.

adaptation — (1) the adjustment of an organism to its environment, (2) a structural or physiological feature which increases an organism's chances of surviving in its **habitat**.

Rhizogonium dozyanum **acute** single-celled marginal teeth. 50 µm

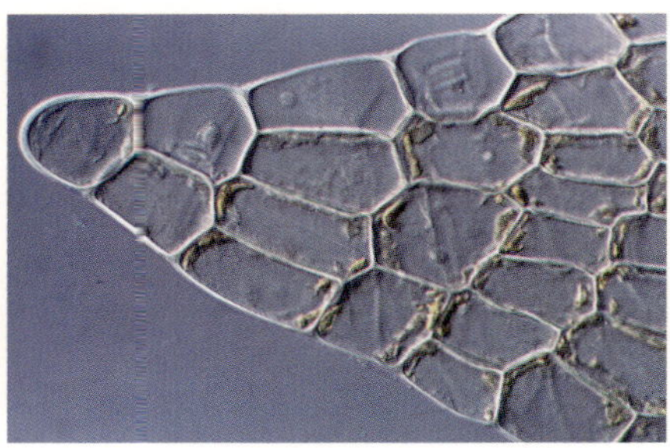

Cephalozia bicuspidata **acute** apex of leaf lobe. 50 µm

adaxial — facing toward a stem or an **axis**, hence a leaf's "upperside" or "inner" surface (*the opposite of* **abaxial**).

adhering (or **adherent**) — (1) sticking to an organ of a different kind (*the opposite of* **cohering**, sticking to an organ of the same kind) (*compare with* **adnate**, fused rather than sticking to an organ), (2) firmly attached to a **substratum** (•below).

Austrolejeunea olgae **adhering** to leaf substratum. 1 mm

Macromitrium longipes **adhering** to rock substratum. 1 cm

adnate — fused to a structure or organ of a different kind (•below, left and middle) (*the opposite of* **connate**, fused to a structure or organ of the same kind) (*compare with* **adhering**, sticking to an organ of a different kind, and **cohering**, sticking to an organ of the same kind).

adsorption — the concentration of molecules or ions (usually a gas or liquid) at an interface or on the surface of a solid.

aduncous — shaped like a hook (•below, right) (*synonyms* **hamate**, **hooked**, and **uncinate**).

Dendroligotrichum dendroides peristome teeth **adnate** to epiphragm. 1 mm

Chiloscyphus fissistipus underleaves **adnate** to lateral leaves both sides. 1 mm

Ptilium crista-castrensis **aduncous** stem tip. 1 mm

adventitious — produced in an atypical location, for example **rhizoids** on a **costa** or on the surface of a leaf (•below, left).

adventive — (1) said of an introduced **species** (*the opposite of* **indigenous**) (•below, middle and right), (2) a branch or some other structure growing from a **dedifferentiated** cell.

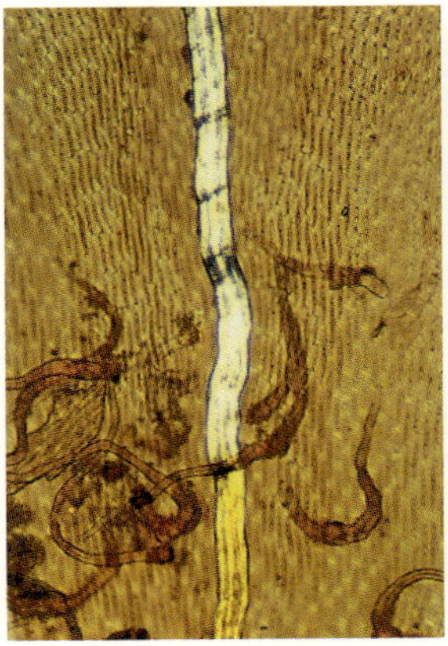

Dicnemon calycinum leaf with **adventitious** rhizoids on upper surface. 100 μm

Fissidens taxifolius, a native of the Northern Hemisphere now **adventive** in the Southern Hemisphere. 1 cm

Campylopus introflexus, an austral moss recently **adventive** in Europe and America. 1 mm

adverse — directly opposite.

aerenchyma — a tissue with large air spaces formed by the invagination of external surfaces and so ultimately connected to the exterior. In oxygen-poor (**anaerobic**) **habitats**, aerenchyma can enhance the diffusion of oxygen to cells.

aerobic — (1) said of a **habitat** that has free oxygen (*the opposite of* **anaerobic**, said of a habitat that lacks free oxygen), (2) said of an organism that uses molecular oxygen as an electron acceptor in its cellular respiration (*the opposite of* **anaerobic**, said of an organism that uses an electron acceptor other than molecular oxygen in its cellular respiration).

aerophyte — an **epiphyte** that grows on the above-ground parts of another plant.

aeruginose — bluish-green, the color of weathered copper (*similar to* **verdigris**).

aff. (abbreviation for **affinis**) — Latin word meaning *related to*, used before the **specific epithet** (**species** name) of a specimen that hasn't yet been identified with certainty (e.g. *Bryum* aff. *billardierei*).

affinity — genetic, **taxonomic**, or geographical relationship.

aggregate — clustered together (•below) (*compare with* **coacervate**, heaped or piled together).

Campylopus clavatus **aggregate** capsules. 1 cm

Dicranum plurisetum **aggregate** capsules. 1 mm

air-chamber — in some marchantialean liverworts, a **pored** cavity in the **thallus** which often is lined with **photosynthetic filaments** (•below, left).

air-pores — openings to **air-chambers** in the upper surface of some marchantialean liverworts (•below, right) (*see* **simple** and **compound pore**).

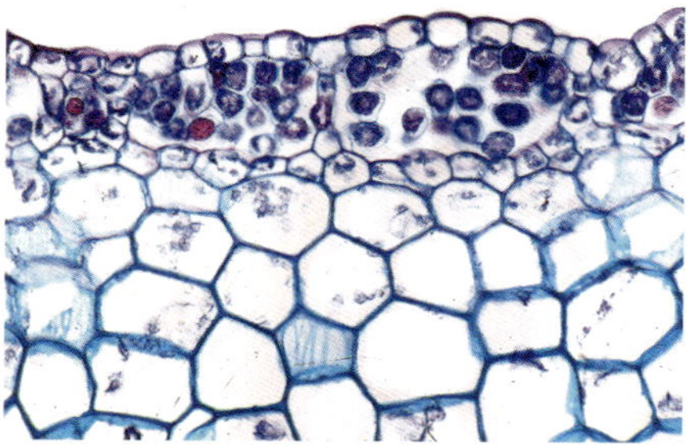

Marchantia sp. **air-chamber**. 100 μm

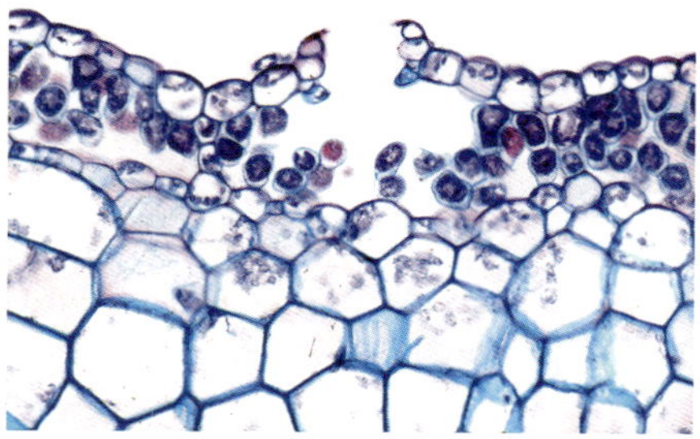

Marchantia sp. **air-pore**. 100 μm

alar cells — specialized cells at the basal corners of a moss leaf which differ from other leaf cells in their size, shape, color, or the thickness and ornamentation of their walls (•below) (*see* **alar region**).

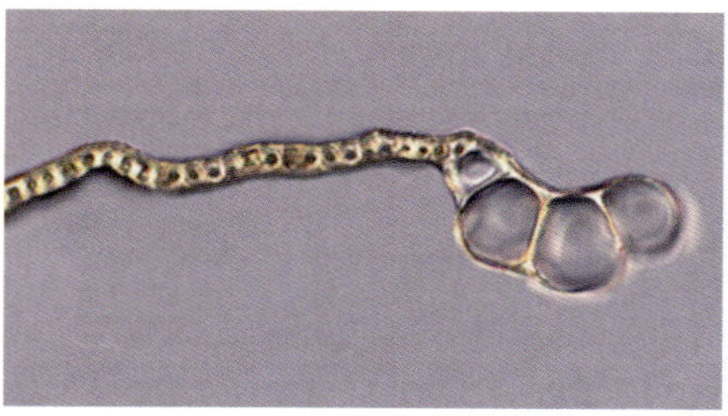

Breutelia pendula leaf xs, inflated **alar cells**. 100 μm

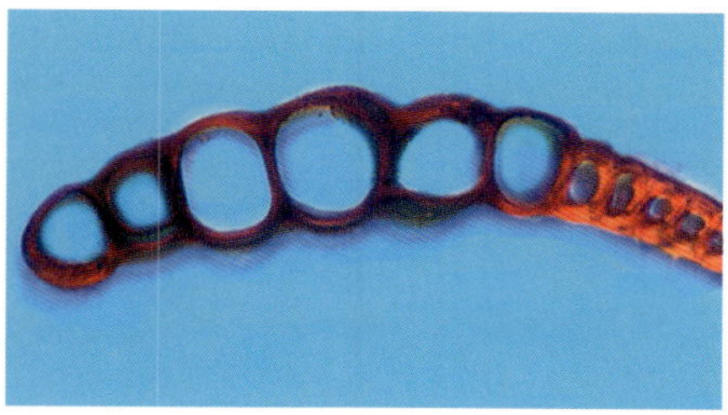

Dicranum plurisetum leaf xs, **alar cells**. 100 μm

alar region — an area (typically triangular) at the basal corner of a moss leaf which controls the orientation of the leaf in response to changes in moisture. Cells of the alar region of many mosses are thin-walled and inflated. They're **turgid** when the leaf is fully hydrated, but collapse when they lose water during dry periods or droughts. Their collapse pulls the leaf closer to the stem, thus minimizing further water loss (•below) (*see* **alar cells**).

Breutelia pendula leaf wm, **alar region**. 100 µm

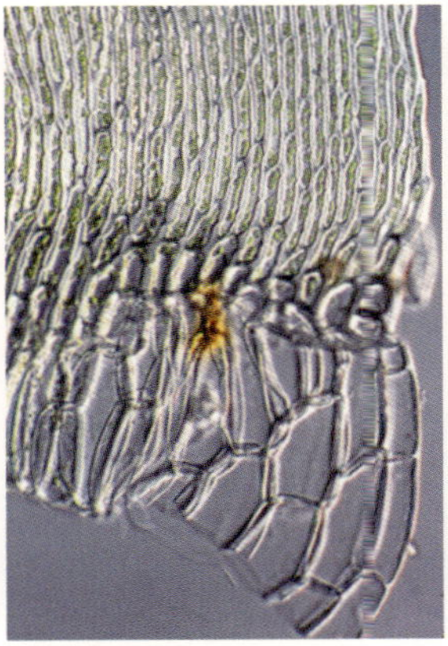

Wijkia extenuata leaf wm, **alar region**. 100 µm

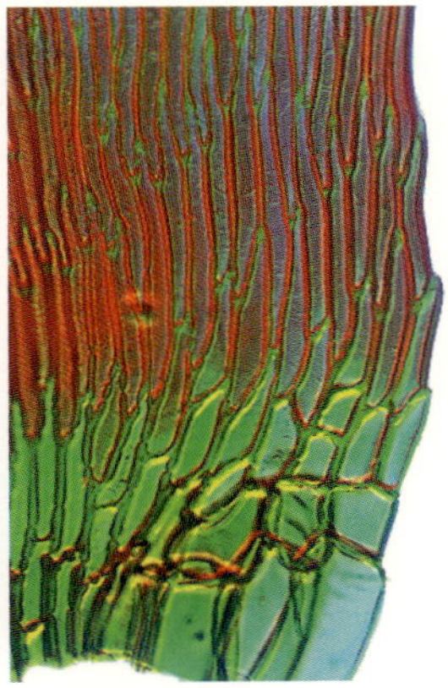

Brotherella henonii leaf wm, **alar region**. 100 µm

alate — **winged** (•below, left).

albescent — changing to whitish from some other color upon drying, maturing, or aging (•below, middle and right).

Pallavicinia xiphoides **alate** frond. 1 mm

Leucobryum glaucum **albescent** habit (dry). 1 cm

Leucobryum glaucum **albescent** habit (dry). 5 mm

alleles — variant forms of a **gene** at any particular site (locus) on a **chromosome**.

allopatric (noun **allopatry**) — said of two or more **species** inhabiting entirely separate areas (*the opposite of* **sympatric**, inhabiting the same or overlapping areas).

allopolyploid — a **polyploid** in which one or more sets of **chromosomes** came from different **species** (*compare with* **autopolyploid** in which all of the chromosomes came from the same species) (*see also* **polyploid** and **ploidy level**).

alluvium (plural **alluvia**) — mineral-rich soil that was deposited by water.

alpestrine — referring to high mountains or regions above timberline.

alpine — the zone of a mountain above timberline.

alternate — (1) said of leaves or branches that are positioned along a stem first on one side and then the other (•below) (*the opposite of* **opposite**), (2) said of a **peristome** with **exostome teeth** that alternate with the **endostome segments**.

Orthorrhynchium elegans **alternate** leaves.
0.5 mm

Acrobolbus lophocoleoides **alternate** leaves.
1 mm

Pseudolophocolea denticulata **alternate** leaves.
1 mm

alternation of generations — a reproductive cycle in which two types of **multicellular** plants give rise to each other in a continuous cycle. One type is **haploid** (each of its cells contains only a single set of **chromosomes**, symbolized by 1n). By **mitosis** it produces **gametes** (male and female sex cells)—hence it's called a **gametophyte**. Two gametes fuse to form a **zygote**, which is **diploid** because it contains two sets of chromosomes (symbolized by 2n), one from each gamete. The zygote then germinates and grows into the other type of plant in the cycle, which by **meiosis** (reduction-division) produces **spores**—hence it's called a **sporophyte**. Each spore germinates and grows into a new gametophyte, thus completing one turn of the cycle.

Psilopilum crispulum still-photosynthetic capsules of young sporophytes.
2 mm

alternation of generations (continued)

Most plants have such a reproductive cycle, but usually one of their two alternating generations is smaller, simpler, shorter-lived, and at least partly parasitic on the other. In the bryophytes, the gametophyte is the dominant generation—it's larger, more complex, longer-lived, and provides the bulk of the sporophyte's water, mineral nutrients, and energy, even though most moss sporophytes contain some chlorophyll and are **photosynthetic** at least when young • •left and below).

As it happens, the two generations don't always alternate, because both can bud off cells which develop into plants which at least *look like* the other generation (*see* **apogamy** and **apospory**).

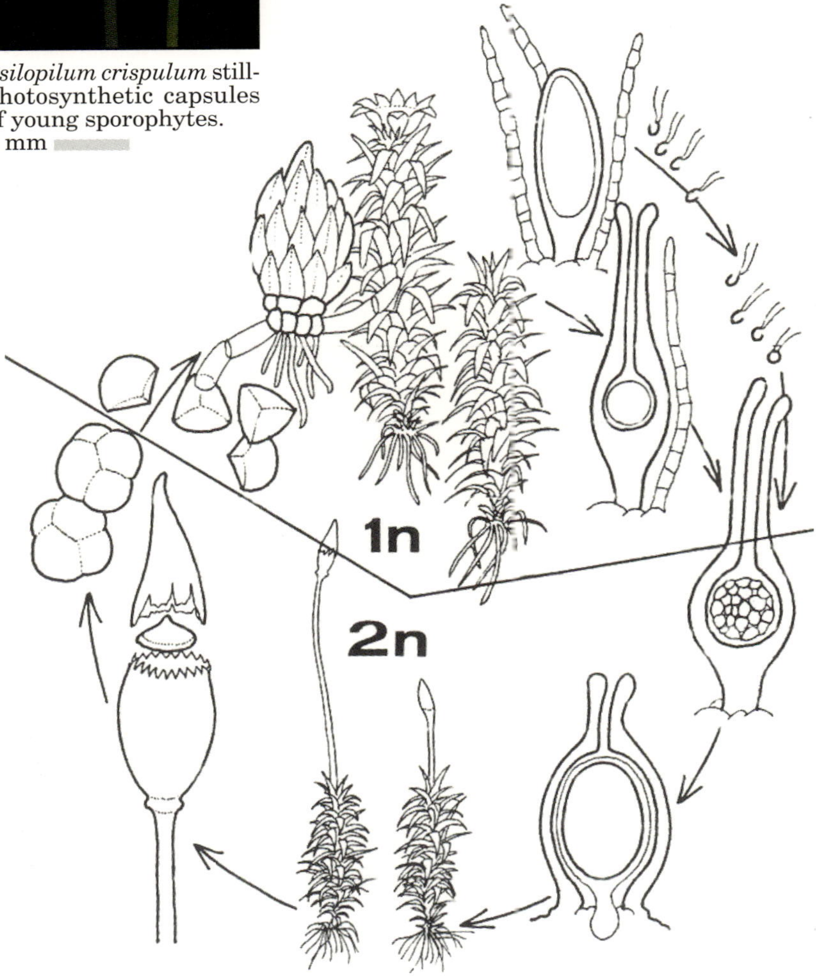

alveolate — having honeycomb-like depressions on the surface, as in some moss **spores**.
amber — the fossilized resin of ancient conifers. Amber often contains well-preserved insects, a fact pivotal to the plot of the science-fiction novel *Jurassic Park*. Exquisitely preserved mosses and liverworts have also been found in amber, providing some of the best evidence for reconstructing the former distribution of modern bryophyte genera.
amentiform — resembling a catkin.
amentulum (plural **amentula**) — the "catkin" of the male **androecium** in *Sphagnum*.
amorphous — formless or nearly so, lacking any definite shape.
amphi- (or **ampho-**) — prefix meaning *on both sides*, *double*, or *around* (Greek, *synonym* in Latin **ambi-**).
amphigastria (singular **amphigastrium**) — leaves, typically in a row on the lower (**ventral**) side of a stem, and usually smaller and with a shape different from that of the other leaves on the stem (•below, and facing page) (*synonyms* in the leafy liverworts **underleaves** and **ventral leaves**).

Cyathophorum bulbosum **amphigastria**. 1 mm

Hypopterygium rotulatum **amphigastria** (UV epifluorescence microscope view). 1 mm

Saccogynidium decurvum **amphigastria**. 1 mm

amphigenous — growing all around a structure.
amphithecium (plural **amphithecia**) — the outermost layer of tissue in an embryonic **capsule**. The amphithecium surrounds the **endothecium**, and gives rise to the outer **jacket** of the capsule in all bryophytes, and in *Sphagnum* and the hornworts gives rise as well to the **spore-sac** of the capsule.
amphitropic — said of anything that curves around the structure it's attached to.
amplectant — embracing, clasping by the base.

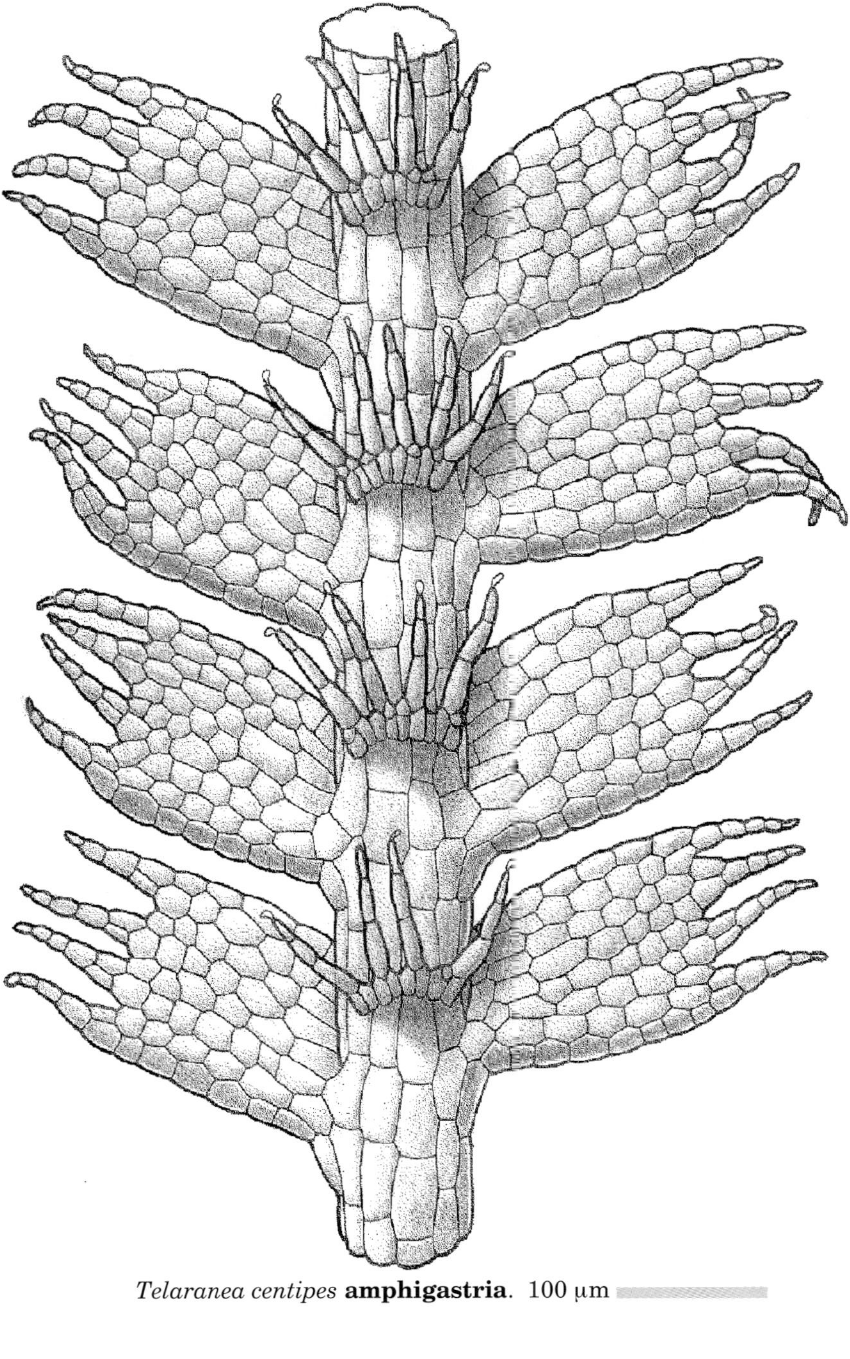

Telaranea centipes **amphigastria**. 100 μm

amplexicaul — a structure that clasps a stem, such as the base of a leaf (•below, left and middle).

ampliate — enlarged or increased (•below, right).

Dawsonia superba **amplexicauls**. 1 mm

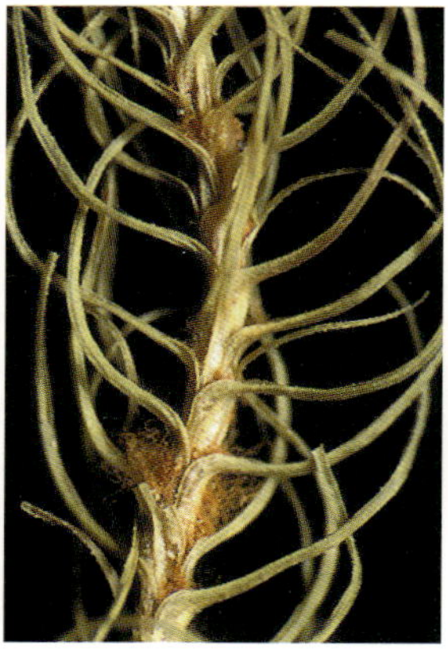

Bartramia pomiformis var. *elongata* **amplexicauls**. 1 mm

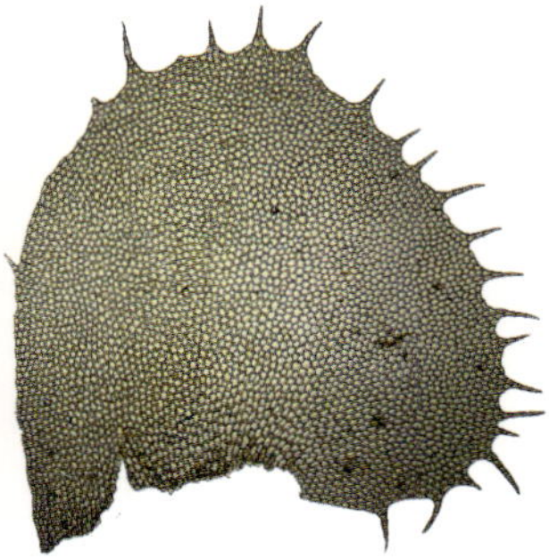

Plagiochila sinclairii **ampliate** leaf base. 0.5 mm

ana- — prefix meaning *again, against, back,* or *backward* (*synonym* **re-**).

anacrogyny (adjective **anacrogynous**) — the production of **archegonia** from the **lateral** or surface cells of a **shoot** rather than the **apical** cell, a type of growth which spares the apical cell and therefore allows further **vegetative** growth of the shoot (*compare with* **acrogynous**, the production of archegonia from an apical cell). **Thallose** liverworts typically are anacrogynous, whereas most leafy liverworts are acrogynous.

A *Sphagnum* bog, an **anaerobic** habitat.

anaerobic — (1) said of a **habitat** that's lacking in free oxygen, such as a *Sphagnum*-dominated **bog** (•left) (*the opposite of* **aerobic**, said of a habitat that has free oxygen), (2) said of an organism that ordinarily uses an electron acceptor other than molecular oxygen in its cellular respiration (*the opposite of* **aerobic**, said of an organism that does use molecular oxygen as an electron acceptor in its cellular respiration).

analogous (noun **analogy**) — said of two structures or organs that have the same function but a different **phylogenetic** origin (*the opposite of* **homologous**).

anastomosing — joined into a network.

androcyte — the cellular precursor of a **spermatozoid** (**antherozoid**).

androcyte mother cell — a **precursor** cell that divides to form two **androcytes**, which in turn develop into **spermatozoids**.

androecial branch — a specialized branch bearing male sex organs (**antheridia**).

androecium — literally the "male house", hence all of the **antheridia** (and also **paraphyses** if present), plus any surrounding leaves (which often are modified) (•below, left).

androgenous — bearing **androecia** (•below, middle and right).

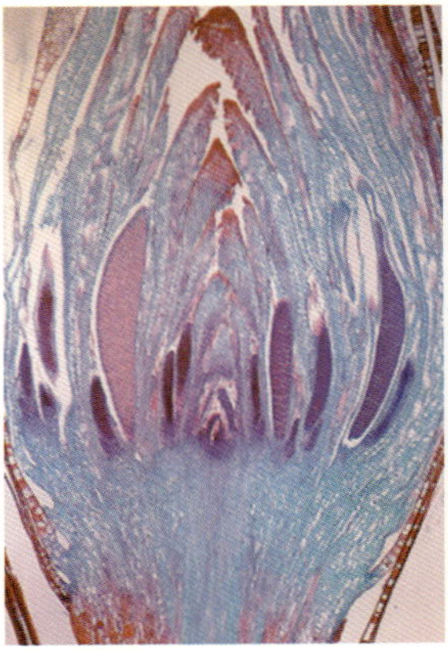

Polytrichum sp. **androecium** (longi-section). 100 µm

Catharomnion ciliatum **androgenous** branch. 1 mm

Tylimanthus saccatus **androgenous** branch. 5 mm

androgonial cell — the **initial** cell of a developing **antheridium** that gives rise to **androcyte mother cells**, then **androcytes** and **antherozoids**.

androgynous — **monoicous**, producing both male and female sex organs (**antheridia** and **archegonia**) in the same **inflorescence** and not grouped or clustered by sex (*synonym* **synoicous**).

angular cells — the group of cells at the basal angles of a leaf.

angulate — angled or having angles or corners.

angustate — (1) narrow, (2) narrowing with **concave** sides.

anisophyllous — (1) producing two types of leaves on the same stem (•left), (2) producing distinctly different stem and branch leaves (*the opposite of* **isophyllous**).

Racopilum convolutaceum **anisophyllous** leaves. 1 mm

anisoplicate — having pleats, folds, grooves, or furrows which are different shapes or sizes (*the opposite of* **isoplicate**, having similar pleats).

anisosporous — producing **spores** of two sizes in the same **capsule** (*not the same as* **heterospory** in the higher plants, which also produce spores of two sizes but in different structures).

anisospory — the production of **spores** of two sizes in the same **sporangium**.

annotinous — showing evidence of **annual** growth, such as by **tufts** of leaves crowded on the stem.

annual — a plant that completes its life cycle within a single year or growing season and then dies (*compare with* **perennial**, a plant that lives for several years or takes three or more years to mature). The **sporophytes** of most bryophytes are annuals, whereas the **gametophytes** are **perennials** (•below, left).

annular — formed or shaped like a ring (•below, right) (*see also* **annulus**).

Hypoptygerium commutatum **annual** sporophytes growing from perennial gametophytes. 5 mm

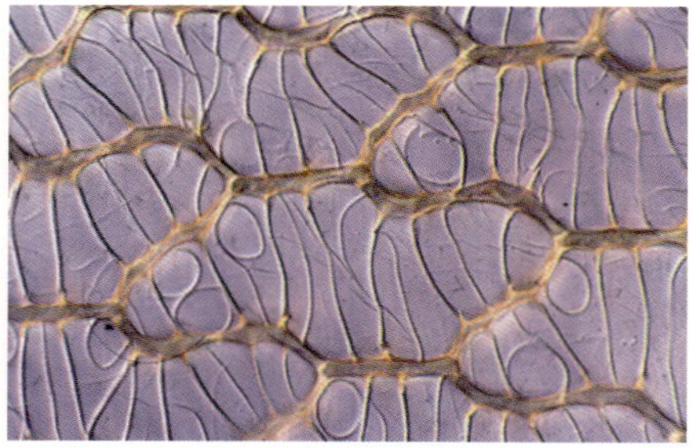

Sphagnum palustre **annular** fibrils of leucocysts. 100 μm

annulate — ringed or surrounded by bands.

annulus — one or more rings of enlarged, specialized elastic cells at the mouth of the **capsule** of most mosses, which assist in shedding the **operculum** (•below, left and middle). The annulus breaks up or peels off during **dehiscence**.

antheridiophore — in some marchantialean liverworts, a stalk that bears **antheridia** at its tip (•below, right) (*compare with* **archegoniophore**, a stalk that bears **archegonia**).

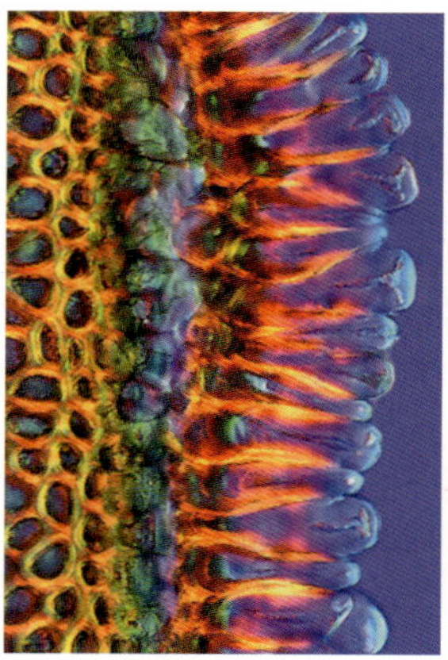

Aulacomnium heterostichum **annulus** (attached). 50 μm

Funaria hygrometrica **annulus** fragment (detached). 50 μm

Neohodgsonia mirabilis **antheridiophores**. 5 mm

antheridium (plural **antheridia**) — a bryophyte's male reproductive organ. When mature, it's a sac containing many **sperms** (**antherozoids**) in a single-layered **jacket** on a short stalk (•below).

Catharomnion ciliatum immature **antheridia**. 100 µm

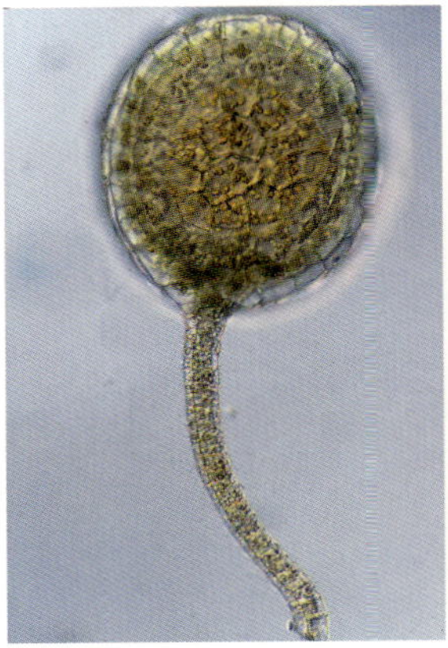

Tylimanthus saccatus **antheridium**. 0.5 mm

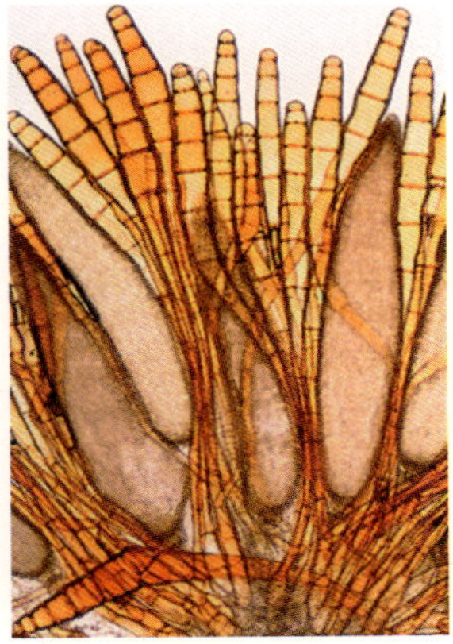

Breutelia pendula sterile paraphyses and immature **antheridia**. 100 µm

antherozoid — a **biflagellate sperm** (male **gamete**) of a bryophyte (•below, left and middle) (*synonym* **spermatozoid**).

anthocyanin — a water-soluble blue, purple, or red pigment restricted to the vacuoles of plant cells.

anthracine — coal-black (•below, right).

anthropochorous — associated with humans or dispersed by humans.

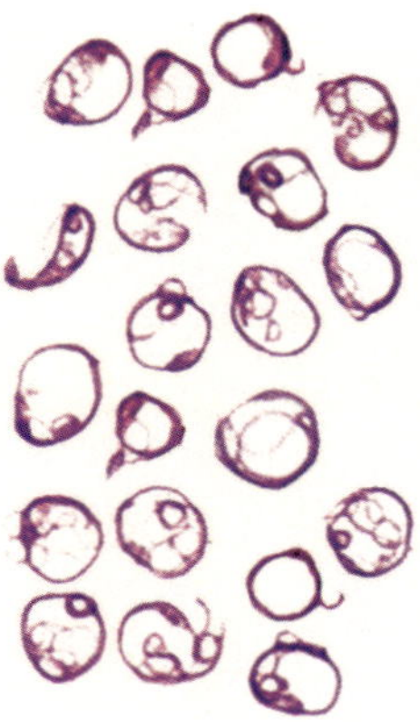

Moss **antherozoids**. 10 µm

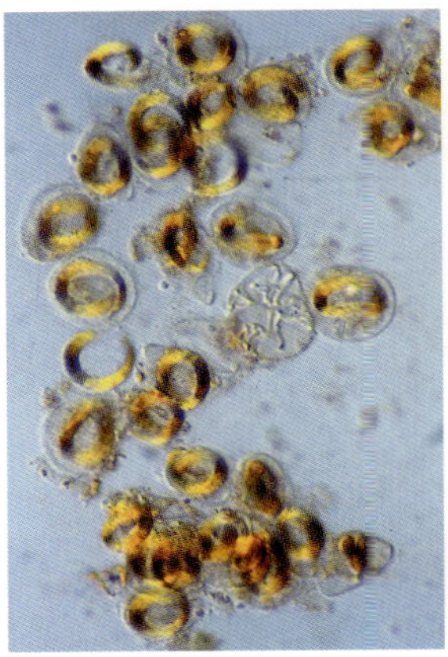

Tylimanthus saccatus **antherozoids**. 10 µm

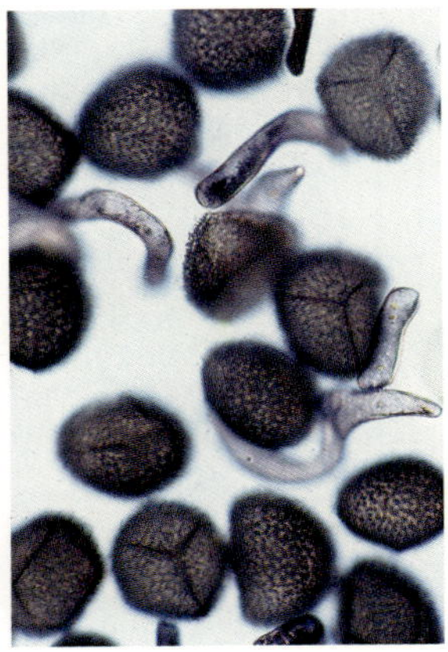

Anthoceros laminiferus **anthracine** spores. 50 µm

antical — (1) facing toward the apex (*the opposite of* **postical**, facing toward the base), (2) the side of a **thallus** or a stem that faces away from the **substratum**.

anticlinal — oriented perpendicular to the surface (*the opposite of* **periclinal**, oriented parallel to the surface).

antrorse — facing or pointing forward or toward the tip, for example **papillae**, **cilia**, or spines on a leaf or stem (•below, and facing page) (*the opposite of* **retrorse**, facing backward or away from the tip) (*compare with* **introrse**, facing inward or toward the center, and **extrorse**, facing outward or away from the center).

Pyrrhobryum bifarium **antrorse** marginal teeth. 100 µm

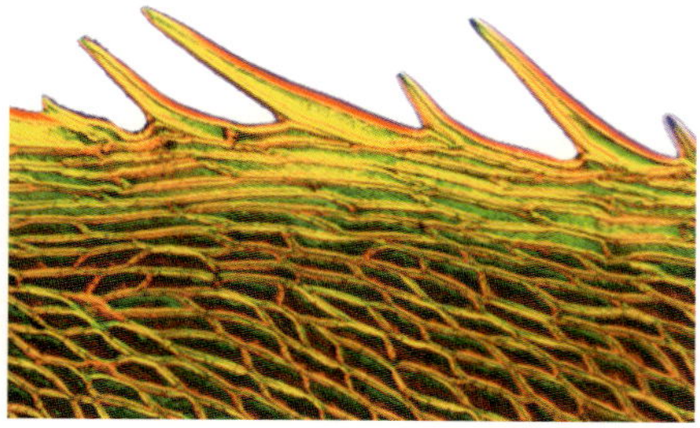

Pseudobryum speciosum **antrorse** marginal spines. 100 µm

aperture — an opening or hole.

apex (plural **apices**) — the tip, point, or summit of a structure such as a stem or leaf (•below).

apical — at or attached to an **apex** or tip.

apical cell — a cell at the tip of the **shoot** which divides to produces leaves and other stem tissues.

Cinclidium stygium leaf **apex** abruptly narrowed to apiculus. 100 µm

Kindbergia praelonga leaf **apex**. 100 µm

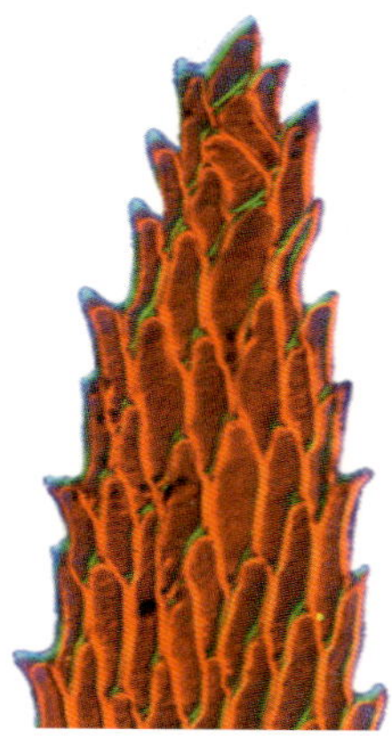

Rhytidiadelphus japonicus leaf **apex**. 50 µm

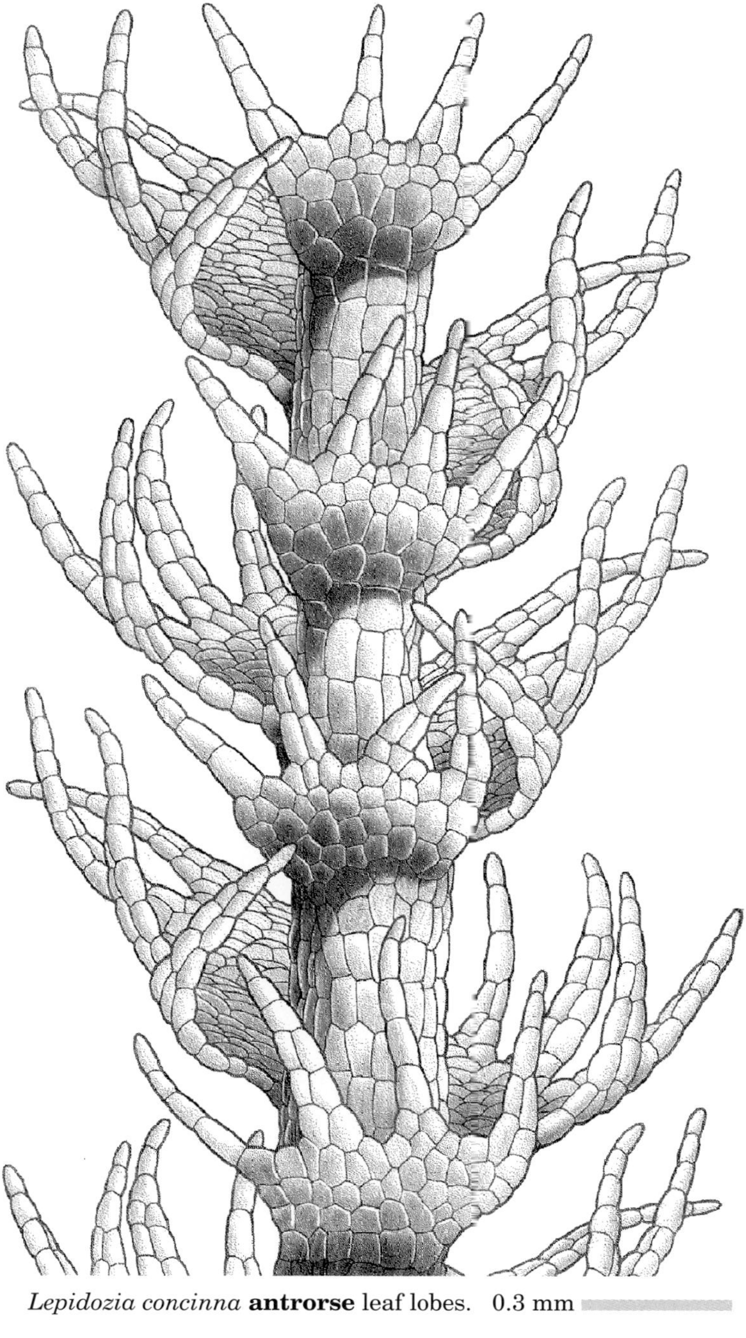

Lepidozia concinna **antrorse** leaf lobes. 0.3 mm

apiculate — (1) in mosses, ending in a short point that's part of the leaf **lamina** rather than the tip of the **costa** (•below), (2) ending in a short, **abrupt** point (**apiculus**) which is longer than a **mucro** (adjective **mucronate**) but shorter and less stout than a **cusp** (adjective **cuspidate**).

Vesicularia ferriei **apiculate** leaf apex. 100 μm

Climacium dendroides **apiculate** leaf apex. 100 μm

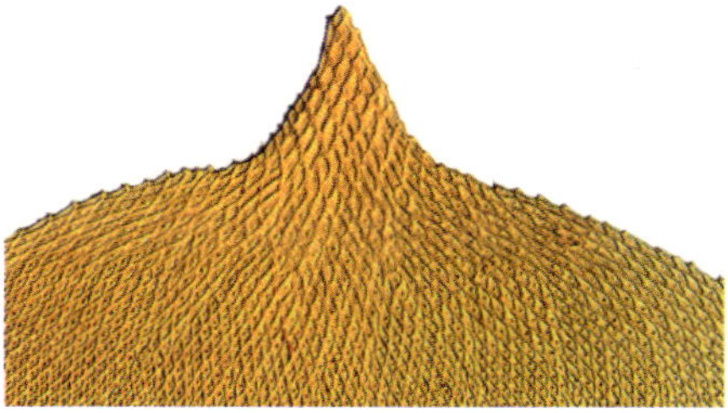

Meteorium buchananii ssp. *helminthocladulum* **apiculate** leaf apex. 100 μm

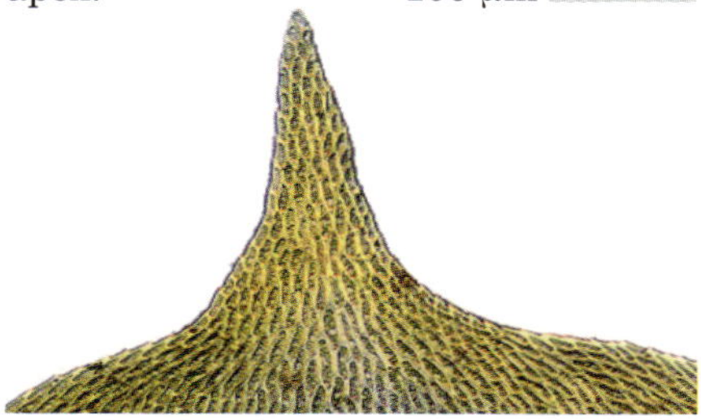

Camptochaete deflexa **apiculate** leaf apex. 100 μm

apiculus (plural **apiculi**) — a short **abrupt** point at a tip or **apex** (•below).

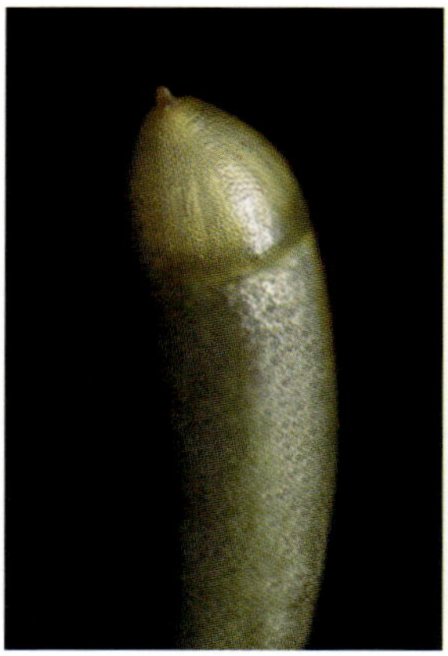

Camptochaete deflexa operculum with **apiculus**. 1 mm

Pleurophascum grandiglobum var. *decurrens* capsule **apiculi**. 1 cm

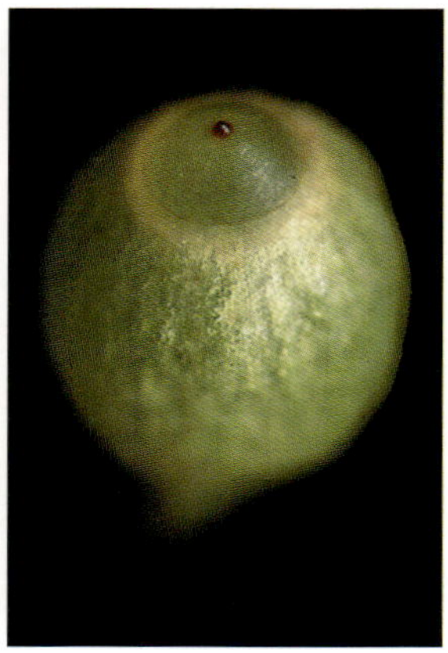

Breutelia pendula operculum with **apiculus**. 1 mm

apo- — prefix meaning *from, away from, free of,* or *out of* (Greek, *synonym* in Latin **ab-** or **abs-**).

apogamy — the formation of a **sporophyte** directly from a **gametophyte** rather than by the fusion of **gametes**—hence the sporophyte has the same number of sets of **chromosomes** in its cells as the gametophyte does, usually one (*see* **apospory**, **haploid**, and **polyploidy**). Apogamy is common among bryophytes.

apolar — lacking a pole, said of marchantialean **spores** in which the **germ tube** ruptures the spore coat at unpredictable sites (*the opposite of* **polar**).

apomixis — (1) in bryophytes, any kind of **vegetative** reproduction, (2) development of a **sporophyte** from an unfertilized **egg** or from the fusion of an egg with a vegetative cell rather than a **sperm**.

apophysis (plural **apophyses**) — in mosses, the **sterile** basal **neck** of a **capsule** where it joins the **seta**, often swollen and with **prominent stomata** (•left and below) (*synonym* **hypophysis**).

Bryum billardierei capsules with **apophyses**.
1 mm

Pyrrhobryum mnioides mature capsules with **apophyses**. 1 mm

apoplastic water — water that's held in the polymers making up the cell walls of a bryophyte (*compare with* **symplastic water**, water held within the cell protoplast or **lumen**, and **capillary water**, water held externally by surface tension).

apospory — formation of a **gametophyte** directly from a **sporophyte** rather than from the **germination** of a **spore** produced by **meiosis**—hence the gametophyte has the same number of sets of **chromosomes** in its cells as the sporophyte does, usually two (*see* **apogamy**, **diploid**, and **polyploidy**). Apospory is common among bryophytes.

appendage — in marchantialean liverworts, a membranous flap that's attached to each ventral **scale** and protects the **apical** cell.

appendiculate — in mosses, having short **transverse** bars or projections, as on the **cilia** of the **inner peristome** (**endostome**).

apple moss — the common name given to the moss family Bartramiaceae because of the fancied likeness of their plump young **capsules** to green apples. Genera of the family include *Bartramia*, *Breutelia* (•left), *Conostomum*, *Philonotis*, and *Plagiopus*.

Breutelia pendula "**apple moss**" immature capsules.
5 mm

appressed — (1) flattened or pressed against the stem or **substratum** (•left), (2) lying close together.

approximate — near or close together but neither touching nor overlapping, as with **peristome teeth** (•below) (*compare with* **contiguous**, touching or overlapping but not fused, and **connivent**, converging at the tips but not fusing).

Bryum argenteum leaves **appressed** against stem. 1 mm

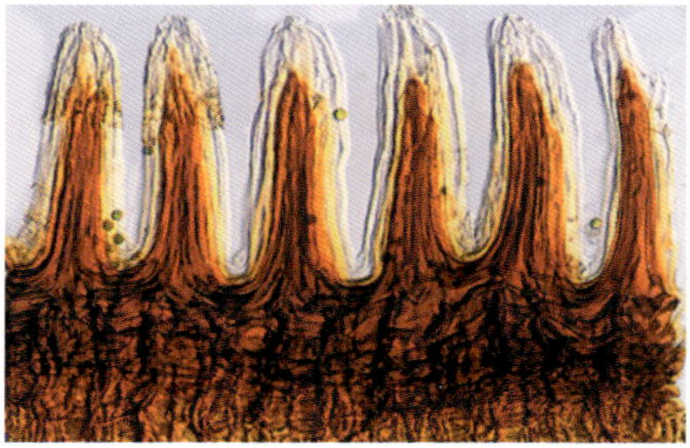

Pogonatum subulatum **approximate** peristome teeth. 100 µm

aquatic — growing or living in water (•below, left).

arachnoid — cobwebby, covered with thin, white, entangled hairs (*see* **hairy** for similar terms).

arboreal — growing in, on, or among trees.

archegoniophore — in marchantialean liverworts, a stalk bearing **archegonia** (•below, middle) (*compare with* **antheridiophore**).

archegonium (plural **archegonia**) — a bryophyte's female reproductive organ. At maturity, it's flask-shaped with a long hollow **neck**, a swollen base (**venter**) containing an **egg** (**ovum**), and a stalk. (•below, right). It protects, nourishes, and influences the growth of the young **sporophyte**.

Blindia lewinskyae **aquatic** habit. 1 mm

Marchantia berteroana **archegoniophore**. 1 mm

Bryum sp. **archegonia** and sterile paraphyses. 100 µm

archesporium (plural **archesporia**) — the mass or layer of cells in an embryonic **sporangium** which later produces **spores** by **meiosis** (*synonym* **sporogenous layer**).

arcuate — curved or bent like a bow or the arc of a circle (•left).

area apicalis — a region toward the tip of a leaf or other structure that's distinctive in some way, such as having cells with thickened walls.

area basalis — a region toward the base of a leaf or other structure that's distinctive in some way, such as having cells with thickened walls.

arenicolous — growing on sand or in sandy **habitats** (•below).

Pyrrhobryum bifarium **arcuate** costae.
1 mm

Didymodon australasiae **arenicolous** habit in coarse sand. 5 mm

areolae (singular **areola**) — small, island-like divisions of a surface resembling the angular cracks in mud or in a network (•below, left).

areolate — divided into **areolae** (•below, middle).

areolation — the pattern of cells in a leaf (•below, right).

Marchantia macropora thallus **areolae**.
1 mm

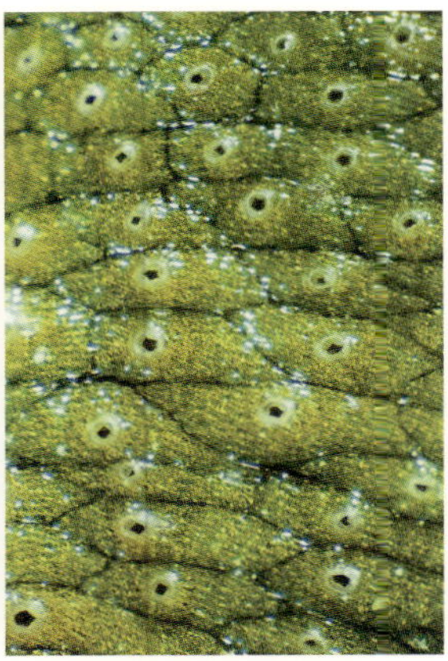

Marchantia foliacea **areolate** thallus surface.
1 mm

Calyptrochaeta cristata **areolation**.
100 μm

argenteous — silvery (•below, left and middle).
argillicolous — growing on clay (•below, right).

Racomitrium lanuginosum **argenteous** hair-points. 5 mm

Bryum argenteum **argenteous** shoots. 5 mm

Fissidens pallidus **argillicolous** shoots. 5 cm

arista (plural **aristae**) — in mosses, a hard, straight **awn** or bristle at the tip of a leaf, usually a projection of the **costa** beyond the end of the leaf blade (•below, left).
aristate — ending in a bristle-point or **awn**, or having a **hair-point** (**arista**) (•below, left).
armature — a general term for spines, **cilia**, or other protective structures (•below, middle and right).
arthrodontous — in mosses, said of a **peristome** that's made up of only parts of cells (*compare with* **nematodontous**, said of a peristome that's made up of *whole* cells).

Leptotheca gaudichaudii **arista**. 50 μm

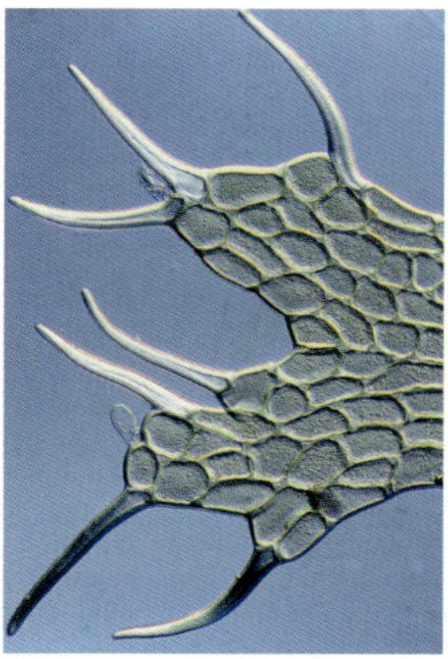

Herzogianthus vaginatus **armature**. 100 μm

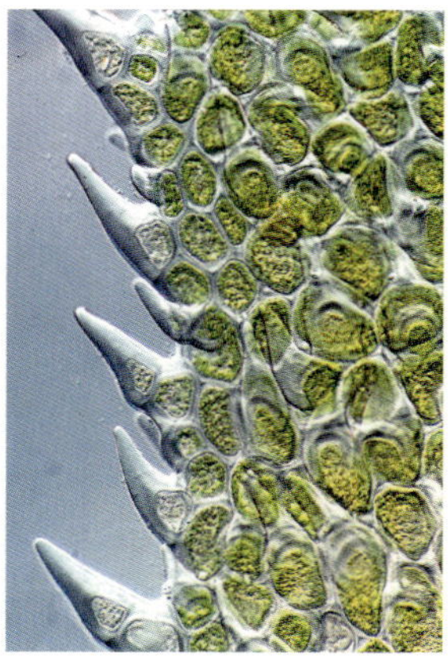

Schistochila muricata **armature**. 100 μm

articulate — (1) having thickened **transverse** lines or joints, as in a **peristome** (•below), (2) said of a structure that's made up of parts fused together at obvious joints.

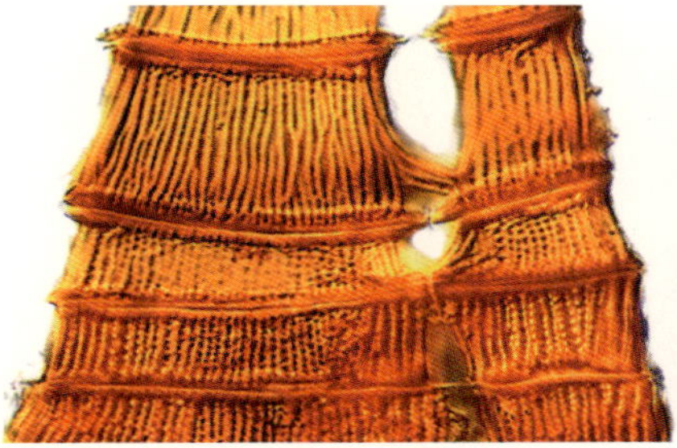

Dicranum platycaulon **articulate** peristome teeth. 50 µm

Campylopus introflexus **articulate** peristome teeth. 100 µm

ascending — sloping or curving upwards (*similar to* **assurgent**).

asexual reproduction — reproduction without the usual **meiosis** or fusion of **gametes**, in mosses by a variety of **brood bodies** (•below), and in liverworts by **gemmae**, **flagelliform branches**, or the **fragmentation** of leaves (•next page).

Achrophyllum dentatum **asexual** brood bodies on leaf margins. 1 mm

Tetraphidopsis pusilla **asexual** brood bodies (wm). 20 µm

Bryum dichotomum **asexual** brood bodies in leaf axils. 10 mm

Campylopus clavatus detached **asexual** brood bodies. 5 mm

asexual reproduction (continued)

Marchantia foliacea **asexual** gemmae cups. 5 mm

Drucella integristipula **asexual** leaf fragmentation. 100 μm

asperate — roughened with projections or stiff hairs.
asperulous — somewhat roughened.
assimilate — energy-rich molecules built from carbon dioxide, mineral nutrients, and the energy of sunlight by the process of photosynthesis plus the many metabolic pathways "downstream" of it (*synonym* **photosynthate**).
assurgent — curving or growing upwards (*similar to* **ascending**).
astomatous — lacking **stomata** (•left) (*the opposite of* **stomatous**).
astomous — in mosses, said of a **capsule** that has no **operculum** or **valves**, and therefore opens (**dehisces**) not along a line of built-in weakness but instead by the decay or breakdown of its wall (*synonyms* **indehiscent** and **cleistocarpous**).
asymmetric (or **asymmetrical**) — lacking formal symmetry, said of anything which can not be cut into two mirror halves along a longitudinal plane that passes through its axis (•below) (*compare with* **radially symmetric** or **actinomorphic** and **bilaterally symmetric** or **zygomorphic**).
athecal — (1) said of branches that are produced at the tip of the stem (*the opposite of* **gyrothecal** and **intercalary**, produced below the tip of the stem), (2) lacking a **sheath**.

Dendroceros validus **astomatous** horn wall. 100 μm

Hypopterygium japonicum **asymmetric** leaf. 0.5 mm

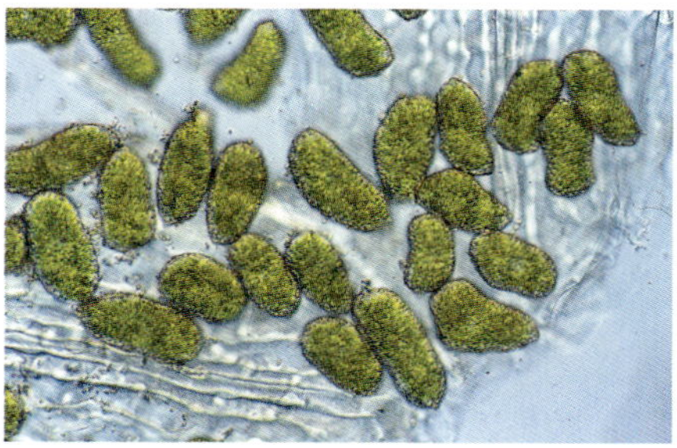

Siphonolejeunea nudipes **asymmetric** spores. 50 μm

atratous — (1) black in part, as at an **apex** (•below, left and middle), (2) turning black.

attenuate — (1) narrowing gradually, (2) tapering gradually to a point (•below, middle and right).

Catharomnion ciliatum **atratous** calyptrae.
1 mm

Ptychomnion aciculare **atratous** and **attenuate** calyptrae.
1 mm

Acrocladium chlamydophyllum **attentuate** tip of vegetative shoot.
1 mm

auricle — an ear-like **lobe** of specialized cells at the base of a leaf near the **margins** (•below, left and middle).

auriculate — with **auricles** (•below, left and middle) (*similar to* **cordate**).

austral — (1) southern, (2) restricted to the Southern Hemisphere (•below, right).

Climacium japonicum **auriculate** leaf base.
100 μm

Ptychanthus striatus **auriculate** leaf base.
100 μm

Pyrrhobryum mnioides **austral** moss.
5 mm

Ptychomnion aciculare **dwarf male** plant.
1 mm

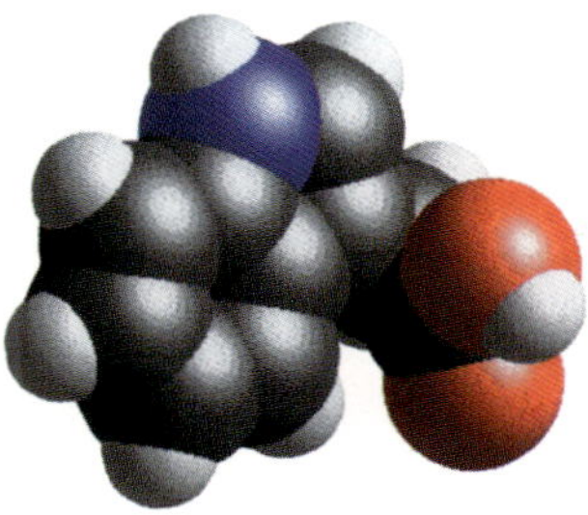

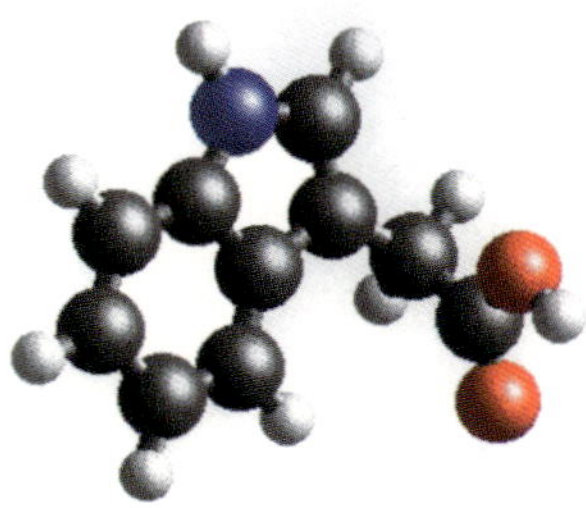

Models of the **IAA** molecule (indole acetic acid): space-filling (upper) and ball-and-stick (lower).

autoicous — **monoicous**, producing both male and female sex organs (**antheridia** and **archegonia**) on the same plant but in separate **inflorescences**. When the male and female inflorescences are on separate branches, the arrangement is said to be **cladautoicous**, whereas if the male is bud-like in an **axil** of the same stem or branch as the female inflorescence, that's called **gonioautoicous**, and when the male plant is a short branch attached to the female with **rhizoids**, that's termed **rhizautoicous**. **Dwarf male** plants growing **epiphytically** on the female (• left) are **dioicous** rather than autoicous, but they're nonetheless sometimes termed **pseudautoicous** or **pseudomonoicous**.

autopolyploid — a **polyploid** in which all of the **chromosomes** came from the same **species**, and usually from a single individual as a result of spontaneous doubling during either **meiosis** (**homologous chromosomes** fail to separate) or **mitosis** (sister **chromatids** fail to separate) (*compare with* **allopolyploid** in which one or more sets of the chromosomes came from different species) (*see also* **polyploid**, **ploidy level** and **nondisjunction**).

auxin — a growth regulator in plants. First studied in vascular plants by Darwin in the 1880s, auxin was named by Frits Went in the 1920s and later was identified as indole acetic acid (**IAA**) (• left). Its many effects depend on its concentration, and include the bending of shoots toward light, growth of fruit, dropping of leaves, and inhibition of lateral buds. Auxin causes its *short*-term effects (such as the bending of shoots toward the light) by controlling the elongation of cells—it merely softens the cell wall, and the cell's internal **turgor pressure** then enlarges the cell. In contrast, auxin causes its *long*-term effects by controlling the expression of **genes**—it regulates gene transcription (the copying of DNA into so-called messenger RNA, which then is transported to ribosomes and translated into protein). Auxin is widely used in agriculture to thin fruit and to prevent premature leafdrop. Synthetic auxins such as 2,4-D are weed-killers.

Auxin is now known to help regulate the growth of mosses. A **filamentous** moss **protonema** has two stages, a **chloronema** followed by a **caulonema**. Only the caulonema produces the tiny **buds** that develop into the moss' familiar leafy **gametophores**, and the switch between the two stages is controlled by auxin. Bud production in turn is controlled by yet another class of plant hormones called **cytokinins**. The shape and size of the **gametophores** are controlled by auxin, cytokinin, and probably calcium ions, which together regulate cell division and cell elongation. Cells which elongate tend to become specialized (they're said to **differen-**

Syntrichia princeps **awn**.
0.5 mm

tiate), whereas cells which divide repeatedly tend to remain unspecialized (embryonic).

Auxin and cytokinins can leak from the protonemata of one moss species and be taken up by those of another species growing nearby—in that way, even unrelated mosses can profoundly affect each other's growth and development (*see* **interactions**).

In marchantialean liverworts, auxin regulates apical dominance and branching patterns, and it's known to control the formation of **gemmae cups**, **rhizoids**, and **underleaves**.

awn — in mosses, a bristle at the tip of a leaf, usually **hyaline** and formed by a **costa**'s projecting beyond the end of the leaf blade (•left).

axial strand — *see* **central strand**.

axil — the angle between a stem and the upper surface of a leaf, branch, or other structure which is growing out of that stem (•below, left).

axillary — said of anything located in an **axil**.

axis (plural **axes**) — (1) a main stem, (2) an imaginary line through the middle of a structure.

baccate — juicy and succulent like a berry.

bar — a unit of atmospheric pressure (the pressure exerted by a column of air on a surface). 1 bar = 100,000 newtons per square meter.

barbate — bearded with **tufts** of long hairs (*see* **hairy** for a comparison of similar terms).

basad — toward or just below the base.

basal — located at the bottom or the near (**proximal**) end of a structure (•below, middle) (*the opposite of* **terminal** or **distal**).

basal membrane — a delicate cylindrical or **tubular** membrane at the base of the **inner peristome** (**endostome**) of the **capsule** of many mosses. It gives rise to **segments** that often **alternate** with **cilia** (•below, right).

Polytrichadelphus magellanicus shoot **axils**.
5 mm

Brotherella henonii leaf **basal** area.
50 μm

Racopilum aristatum **basal membrane**.
100 μm

biomass — the total mass of organisms per unit area.
bipartite — having two parts.
bipinnate — twice-pinnately branched (•below, left) (*compare with* **pinnate**, pinnately branched once).
bipolar — a distribution pattern in which a **species** grows in temperate or polar portions of the Northern and Southern Hemispheres and/or the tropical highlands, but not the tropical lowlands. About 100 species of mosses have a bipolar distribution, ten of them introduced (•below, right).

Hypopterygium filiculaeforme **bipinnate** frond. 1 cm

Campylopus introflexus introduced **bipolar** moss. 1 cm

birefringent — said of a material that splits a single beam of polarized light into two beams which are at right angles to each other and which exit the material out of phase because one is retarded during its passage. The phase differences generate false colors that are clues to the ultrastructure of the material. Until the invention of electron microscopy, birefringence was the most important tool for studying the fine structure of cell walls and cytoplasm. The cell walls of many mosses are birefringent, some dramatically so (•below, left), whereas the cell walls of most liverworts are not.
bisbifid — said of a **bifid** liverwort leaf with **lobes** that are also bifid, but less strongly (•below, right).

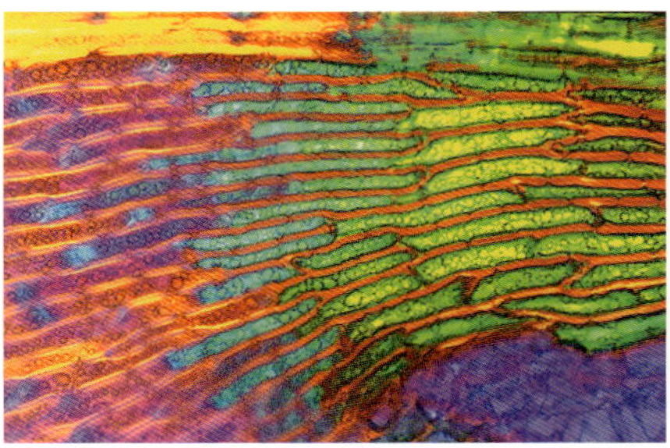

Dicranum plurisetum **birefringent** basal leaf cells. 50 μm

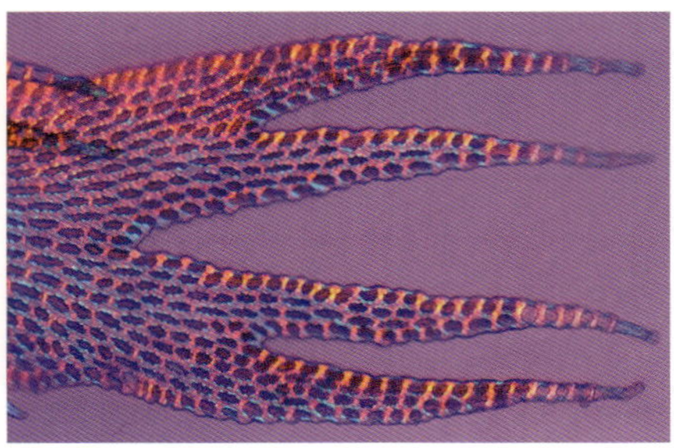

Lepicolea attenuata **bisbifid** leaf. 100 μm

biseriate — in two rows (•left).
bistratose — two cell layers thick (•below, middle).
bivalent — a structure that's formed at the start of **meiosis** by the linkage of two pairs of **chromatids**.
bloom — a waxy or powdery covering that gives a whitish, grayish, or bluish cast to a surface (•below, right) (*synonym* **glaucous**).

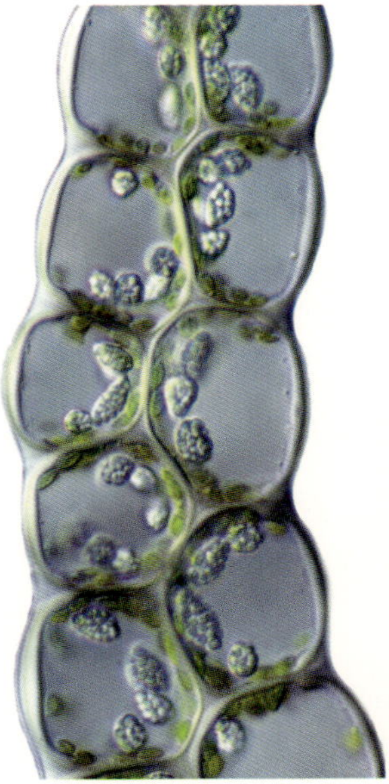

Chiloscyphus semiteres **biseriate** tooth of underleaf. 50 μm

Isolembidium anomalum pluri- to unistratose leaf. 100 μm

Saelania glaucescens surface **bloom**. 1 mm

bog — (1) any waterlogged vegetation, (2) vegetation that's open, **acidic**, waterlogged, dominated by *Sphagnum* mosses and mineral-poor because all of its water comes from rainfall rather than soil or runoff (*synonyms* **ombrotrophic** bog or true bog).
bog iron — deposits of iron in waterlogged vegetation. The deposits have been linked to particular mosses, among them *Brachythecium rivulare*, and are substantial enough to be exploited as iron ore.
bog moss — a common name for *Sphagnum* mosses.
border — in mosses and liverworts, cells along a leaf **margin** that differ from other leaf cells in their shape, size, color, or wall thickening (•below).

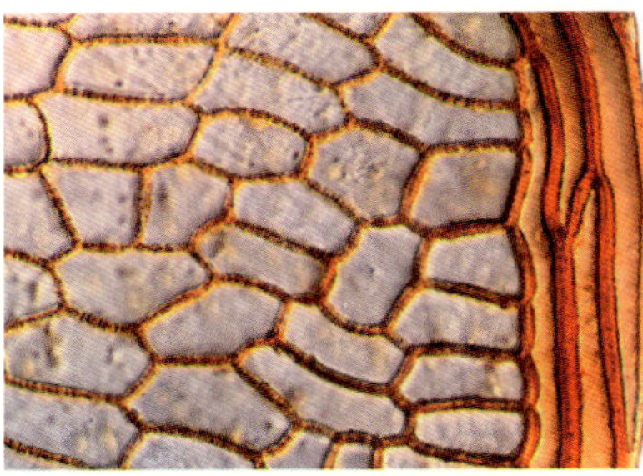

Cinclidium stygium leaf **border**. 50 μm

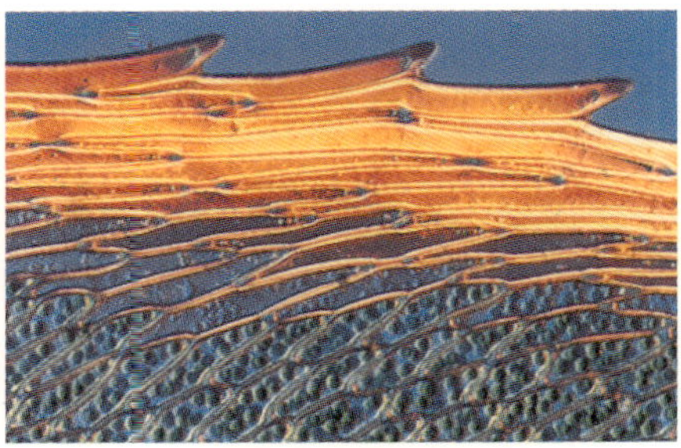

Bryum billardierei leaf **border**. 100 μm

bordered — having a **border** (•below, left).
boreal — (1) northern, (2) restricted to a broad zone of circumpolar coniferous forest.
botryoid (or **botryoidal**) — resembling a bunch of grapes (•below, right).

Bryum billardierei **bordered** leaves. 5 mm

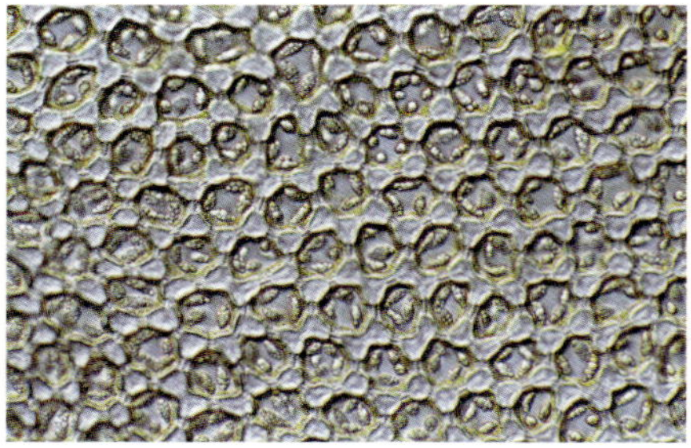

Plagiochila circinalis **botryoid** oil-bodies. 100 μm

bracts — modified leaves surrounding **archegonia**, **antheridia**, or **gemmae cups** (•below).
bracteole — a modified **underleaf**.

Dicranum plurisetum perichaetial **bracts**. 1 mm

Tylimanthus saccatus perigonial **bracts**. 5 mm

brood bodies — any structures that function as **vegetative propagules**, such as **gemmae**, **bulbils**, underground **tubers**, or **reduced** branches (**surculi** or **cladia**) and leaves (•below and next page).

Achrophyllum dentatum **brood bodies**. 200 μm

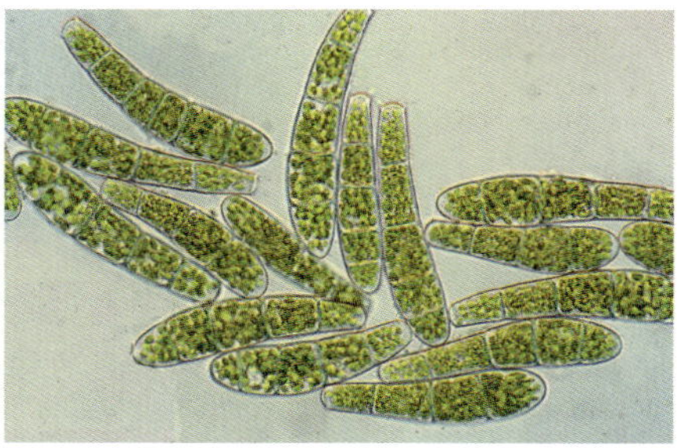

Tetraphidopsis pusilla **brood bodies**. 100 μm

brood bodies (continued)

Bryum dichotomum globose **brood bodies.**
1 mm

Bryum sauteri **brood bodies** (tubers).
100 µm

Leptotheca gaudichaudii **brood bodies** (filaments).
1 mm

broom moss — (1) a common name given to the *Dicranum* mosses and particularly the nearly **cosmopolitan** *Dicranum scoparium* because their leaves are curved strongly toward one side of the stem and so resemble well-used brooms (•below, left) (*see* **falcate-secund**), (2) a common name given to *Pogonatum subulatum* because it was once made into soft brooms for dusting curtains and beds.

brunneus (or **bruneus**) — a general term for brown (•below, right).

Dicranoloma billardierei **broom moss.**
1 mm

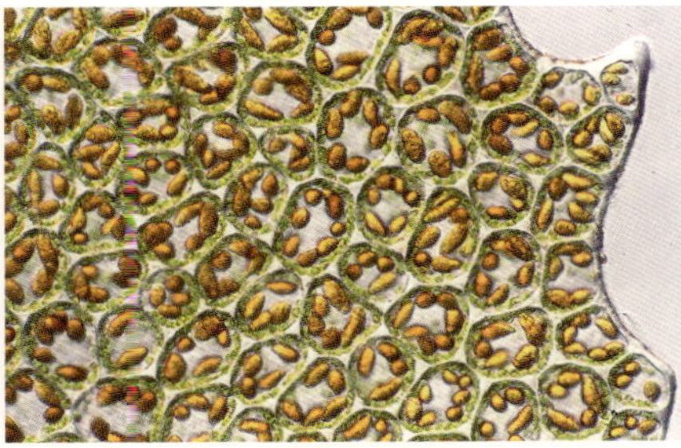

Tylimanthus diversifolius **brunneus** oil-bodies. 100 µm

bryoid — (1) having a *Bryum*-like **peristome** with a well-developed **exostome** and **endostome**, (2) having *Bryum*-like leaves with large, thin-walled **rhombic** or **hexagonal** cells (*see* **hexagonal** for illustration).

bubble cell — a large **alar cell** (*see* **alar cells** for illustration).

bud — a structure produced by a **caulonema**, or a stem which includes an **apical** cell that can develop into a **gametophore** (•below, left).
bulbiform — resembling a flower bulb or onion (•below, right).
bulbil — a tiny **deciduous vegetative propagule**.
bulbous — swollen, bulb-like.

Weissia controversa **bud**. 1 mm

Funaria hygrometrica **bulbiform** shoots.
1 mm

byssaceous — cottony or divided into fine threads.
byssoid — resembling cotton.
caducous — falling off readily or early (*synonym* **deciduous**).
caesious — bluish or with a bluish **bloom** (*compare with* **glaucous**, which is whiter).
caespitose — **tufted**, growing in dense patches, **cushions**, or **turfs**, as opposed to loose **mats**.
calcareous — said of any material containing calcium carbonate, such as limestone or dolomite.
calcicole — an organism largely restricted to **substrata** rich in calcium.
calcicolous — growing on a limy **substratum**.
calcifuge — a plant that's unable to survive in high-calcium **habitats** (*synonym* **calciphobe**) (*the opposite of* **calciphil**, a plant that does best in **calcareous** habitats).
calciphil (or **calciphile**) — a plant that does best in **calcareous habitats** (*the opposite of* **calcifuge** or **calciphobe**).
calciphilous — thriving in **calcareous habitats**.
calciphobe — a plant that's not able to survive in high-calcium **habitats** (*synonym* **calcifuge**) (*the opposite of* **calciphil**, a plant preferring **calcareous** habitats).
callose — (1) hard and thick, (2) pertaining to plugs at the ends of conducting sieve tubes in the phloem tissue of higher plants or in comparable cells in the stems of Polytrichidae mosses.

Racopilum convolutaceum spears with **calyptrae**. 1 mm

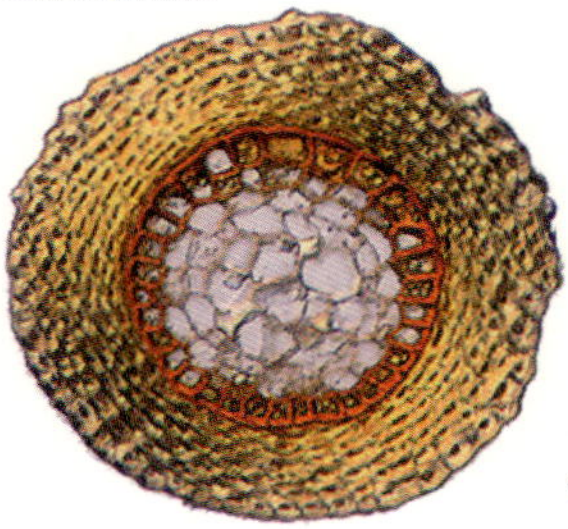

Dicnemon dixonianum xs **calyptra**, sporangium. 100 μm

callus — a mass of thin-walled, unspecialized cells, often formed at a wound.

calyptra (plural **calyptrae**) — in mosses and liverworts, a membranous or hairy cap or hood that forms from the wall of the **archegonium** and protects the embryonic **sporophyte**. It's formed after fertilization of the **egg** from tissues around the base (**venter**) of the **archegonium**, and is ripped off as the **seta** elongates. In most liverworts, it ruptures as the **seta** lengthens, leaving only a **basal** collar (*compare with* **shoot-calyptra**, **stem-calyptra**, or **coelocaule**, an embryo-protecting structure that's derived from stem as well as archegonial tissue and is **fleshy** rather than membranous). In mosses, the collar-like fragment of archegonium tissue that's left behind is called the **vaginula**. A moss calyptra usually falls off just before the **spores** are dispersed. Its function isn't only protection of the developing capsule from abrasion and **desiccation**—as well, it somehow controls development of the capsule itself. There's good experimental evidence for that—if it's slipped off when the sporophyte is young (still in the so-called **spear** stage) (•upper left), the capsule doesn't develop. Its control of the capsule's development is known to be physical rather than chemical, but precisely how it works isn't understood yet (•lower left).

Moss calyptrae come in a dazzling array of textures, shapes, and sizes, and bryologists have exploited them for centuries for classifying and identifying mosses (•below).

campanulate — bell-shaped, like an upside-down cup with a wide-flaring mouth (*compare with* **cupulate** and **cyathiform**, cup-shaped with a mouth that's only slightly flared, and **urceolate**, pitcher-shaped, distinctly narrowed just below the mouth).

Dicranum menziesii capsules with **calyptrae**. 5 mm

Ceratodon purpureus capsules, one with **calyptra**. 5 mm

Racopilum sp. capsules with hairy **calyptrae**. 1 mm

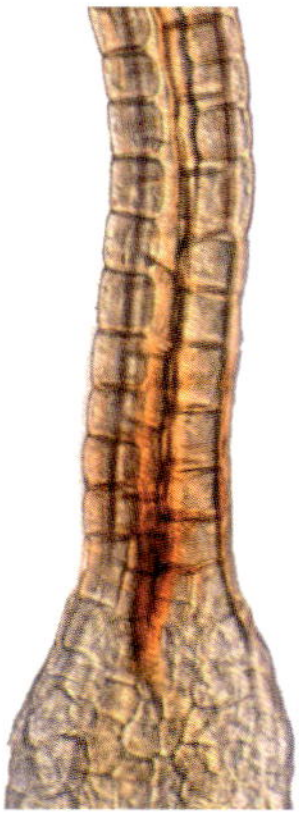

Bryum billardierei **canal cells.** 50 µm

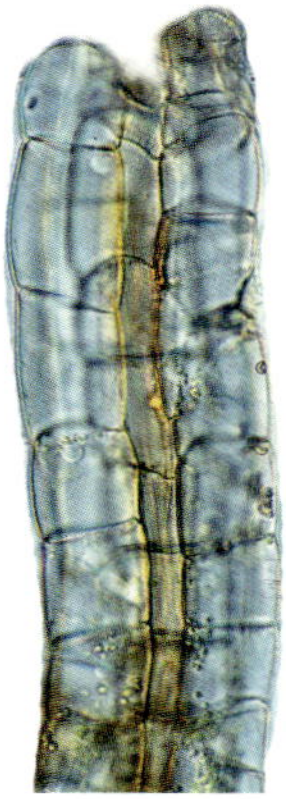

Hymenophyton leptopodum **canal cells.** 50 µm

campestrian — occurring in open country.

canal cells — the central row of cells in the **neck** of an **archegonium** (•upper and middle left). When they eventually disintegrate, they release a sperm attractant into the thin film of water that bathes the archegonium, and by swimming toward an increasing concentration of the attractant, sperms unerringly swim down the neck to the **egg** below.

canaliculate — **grooved** or channeled lengthwise like a canal, gutter, or channel.

cancellate — resembling a lattice or network (•below) (*synonym* **clathrate**).

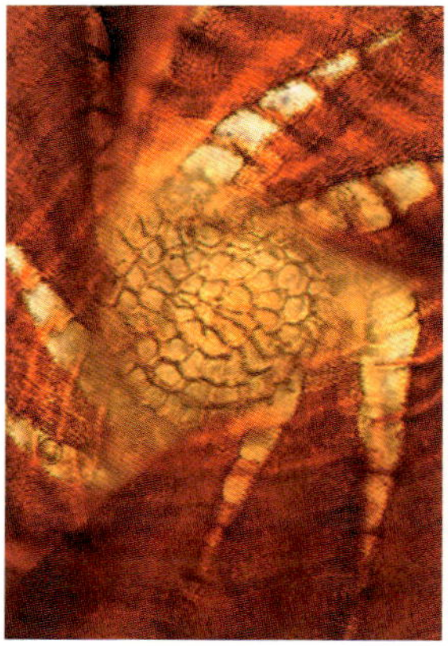

Funaria hygrometrica **cancellate** apical disk of peristome. 100 µm

Fontinalis antipyretica **cancellate** endostome. 100 µm

cancellinae (singular **cancellina**) — the sharply defined, large, empty, and usually **hyaline** cells in the leaf base of mosses of pottioid genera (•below, left) (*synonym* **endohyalocyst**).

canescent — appearing frosted or **hoary** because of numerous **hyaline** hair-points on the leaves (•below, right).

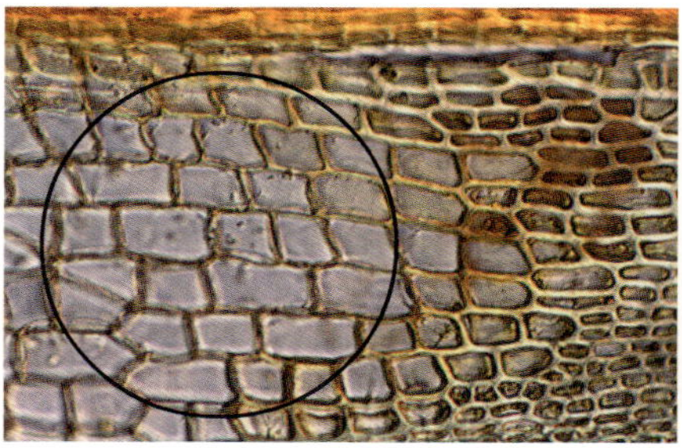

Syrrhopodon japonicus **cancellinae.** 100 µm

Racomitrium pruinosum **canescent** shoot tips. 1 mm

canopy former — a bryophyte with a growth habit that resembles a tree, with the lowest branches on the stem borne well above the **substratum**, thus forming a miniature canopy (•below, left) (*synonym* **dendroid**).

capillary — (1) hair-like (*compare with* **filiform**, thread-like and hence less delicate), (2) a narrow slit or cavity formed between two adjacent leaves, **scales**, or branches (•below, middle).

capillary water — water held externally on the surface by capillarity or surface tension (•below, right) (*compare with* **apoplastic water**, water held in the polymers making up the cell walls, and **symplastic water**, water held within the cell protoplast or **lumen**).

Hypopterygium filiculaeforme **canopy former**.
10 cm

Cyathophorum bulbosum water-filled **capillaries**.
1 mm

Ptychomnion aciculare **capillary water**.
1 mm

capitulum (plural **capitula**) — a head-like mass of crowded branches on the tip of a stem in some bryophytes, for example *Sphagnum* mosses (•below).

Sphagnum sp. **capitula**.
1 cm

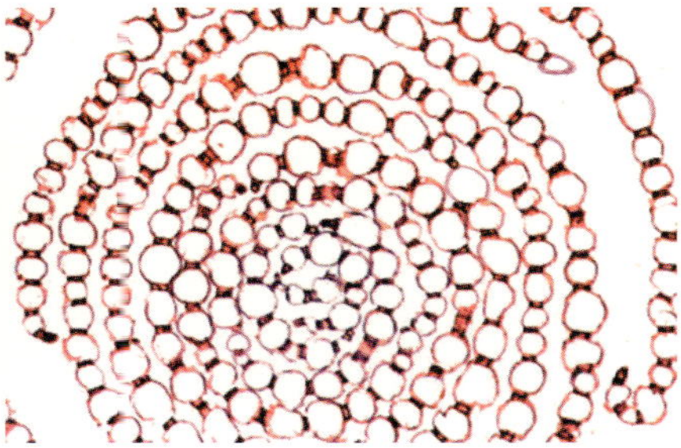

Sphagnum sp. section through leaves of a **capitulum** branch.
100 µm

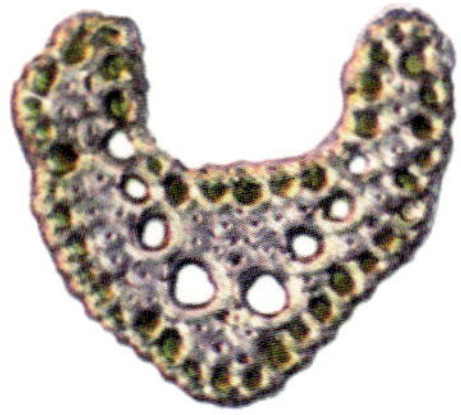

Ditrichum difficile **carinate** leaf (xs).
100 µm

capsule — the **spore**-case (**sporangium**) of a bryophyte, **simple** in liverworts but in most mosses consisting of a sterile base (**neck**, **apophysis**, or **hypophysis**), a **fertile** spore-case (**theca** or **urn**), and usually a sterile **lid** (**operculum**).
carinate — (1) like a boat keel (•left), (2) **winged**.
carneous (or **carneus**) — flesh-colored.
carnose —juicy but firm (•below, left) (*compare with* **fleshy** and **succulent**, which denote a texture that's not so firm).
carpocephalum (plural **carpocephala**) — in some **thallose** liverworts, the **archegoniophore** tip that bears the **sporangia**.
cartilaginous — firm and tough (•below, middle).
castaneous — chestnut-colored (•below, right).

Leucobryum candidum **carnose** leaves.
1 cm

Dicnemon dixonianum **cartilaginous** leaf cells.
100 µm

Hypnodendron comatum **castaneous** capsules.
5 mm

catenulate — resembling a chain (•below, left).
caulescent — having a stem (•below, right) (*the opposite of* **acaulescent**).

50 µm

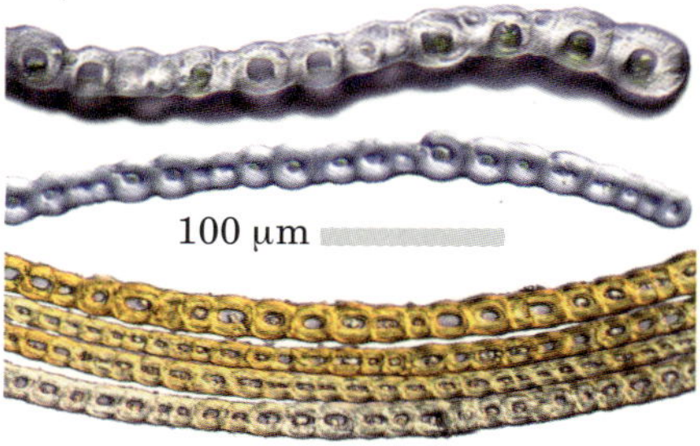

Leaves **catenulate** in cross-section—*Cladomnion ericoides* (top), *Dicnemon dixonianum* (middle), and *Weymouthia cochlearifolia* (bottom).

Triquetrella papillata—**caulescent**.
1 cm

caulidium (plural **caulidia**) — an alternative term for a **haploid** bryophyte stem, preferred by some bryologists (•below, left) (*see also* **phyllid**).

cauline — part of or belonging to the stem.

caulis (plural **caules**) — stem (•below, middle).

caulonema (plural **caulonemata**) — in mosses, the second of the two growth-stages of a typical **protonema** (the first stage is called a **chloronema**). Caulonemata usually have pigmented walls and **oblique** cross-walls, whereas most chloronemata have **hyaline** walls and **transverse** cross-walls. Ordinarily, only the caulonema is able to produce the **buds** which develop into the familiar leafy **gametophores** of an adult moss (•below, right).

Macromitrium longipes **caulidia**.
1 mm

Polytrichadelphus magellanicus **caules**.
1 cm

Weissia controversa chloro- and **caulonemata**.
50 μm

cell — the microscopic structural unit of all living things, in plants consisting of a wall enclosing a cavity (**lumen**) that's in turn bounded by a membrane. Some kinds of bryophyte cells have important functions even after they've died (•below).

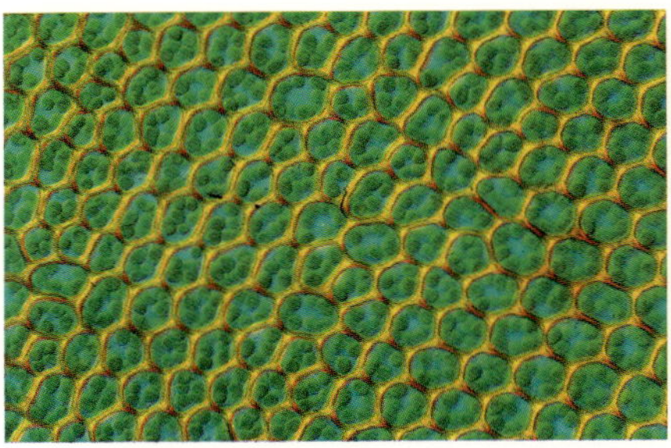

Plagiomnium cuspidatum leaf **cells**.
100 μm

Tetraphis pellucida leaf **cells**.
100 μm

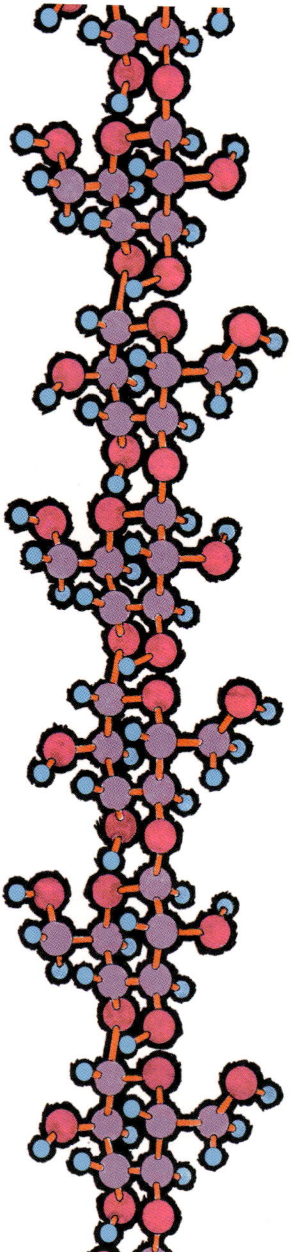

Ball-and-stick model of part of a **cellulose** chain.

cellulose — a complex insoluble carbohydrate consisting of up to 15,000 glucose residues chemically bonded end-to-end. As with most large linear polymers, it has no one size because it's not synthesized on a template. Cellulose makes up the bulk of the cell walls in most plants, including bryophytes, and at any one time, half of the carbon in the biosphere is locked up in it (•left).

central strand (or **axial strand** or **central cylinder**) — a thin column or cylinder of elongate cells at the center of the stem of some bryophytes, best seen in a thin cross-section or a cleared wholemount (•below).

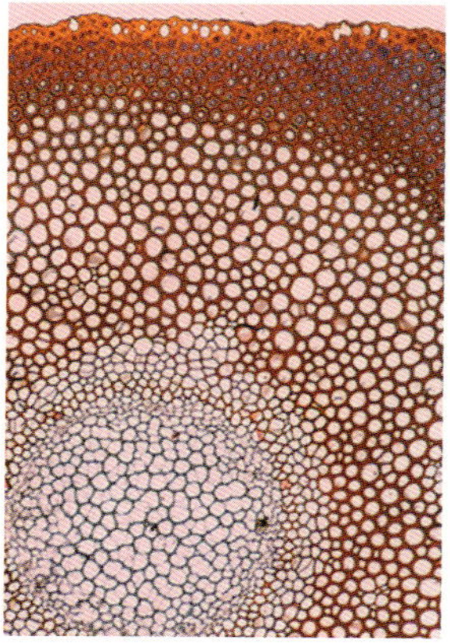

Dendroligotrichum dendroides stem xs, **central strand**. 100 µm

Metzgeria sp. thallus xs, **central strand**. 100 µm

centromere — a **constricted** part of a **chromosome** where two **chromatids** are joined and the microtubules of the spindle attach during **mitosis** and **meiosis**. It's composed of two **kinetochores**, one on each chromatid.

cereus (or **ceraceus**) — waxy.

cernuous — nodding or drooping at ± 45° (•below).

Hypopterygium sp. **cernuous** capsules. 5 mm

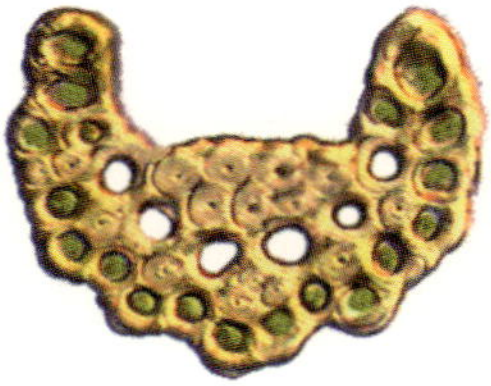

Campylopus clavatus **channeled** leaf (xs).
100 µm

cervine — dark and tawny, the color of a deer.
cf. (abbreviation for **confer**) — a Latin word meaning *compare with*, used either to suggest a possible identity of a specimen which hasn't yet been determined, or to draw attention to the differences and/or similarities among several specimens.
c. fr. (the abbreviation of **cum fructibus**) — Latin phrase meaning *with fruit*, hence producing **sporophytes** or sporulating (•below).
channeled — said of leaves with a longitudinal groove formed by upturned **margins** or (in mosses) a sunken **nerve** (•left) (*synonym* **canaliculate**).

Hypopterygium rotulatum **c. fr.**
5 mm

Bryum billardierei **cum fructibus**.
5 mm

chartaceous — papery (•below, left and middle) (*synonym* **papyraceus**).
chasmocolous — growing in cracks, crannies, or crevices (•below, right).
cheradophilous — preferring a **habitat** of detritus, alluvium, or silt.
chersophilous — preferring dry land **habitats**.

Pulchrinodus inflatus **chartaceous** leaves.
5 mm

Neckera pennata shoot, **chartaceous** leaves.
1 mm

Didymodon australasiae **chasmocolous** habit.
20 cm

chiasma (plural **chiasmata**) — a site where two **homologous chromosomes** have exchanged genetic material during **meiosis**.

chledophilous — preferring muddy, barren, or disturbed **habitats**.

chlorenchyma — a specialized layer of chlorophyll-containing cells forming a richly **photosynthetic** region.

Sphagnum sp. leaf wm, **chlorocysts** and leucocysts.
100 µm

chlorocysts — (1) in a *Sphagnum* leaf, narrow and elongate cells containing chlorophyll and forming a network surrounding larger **hyaline** cells called **leucocysts** (•left) (*synonyms* **chlorophyll cells** and **green cells**), (2) in the leaves of species of the moss family Leucobryaceae, small green cells enclosed by layers of hyaline cells or intermixed with them (•below, left).

chloronema (plural **chloronemata**) — in mosses, the first of the two growth-stages of a typical **protonema** (the second stage is called a **caulonema**). A chloronema typically has **hyaline** walls and **transverse** cross-walls, whereas a caulonema has pigmented walls and **oblique** cross-walls. Only the caulonema ordinarily produces the **buds** which develop into the familiar leafy **gametophytes** of an adult moss.

chlorophyllose — said of a cell that contains chlorophyll and therefore is green, such as the ordinary leaf cells of most bryophytes, the specialized green cells of a *Sphagnum* or *Leucobryum* leaf, as well as some **spores** and the **paraphyses** in immature moss **perigonia** and **perichaetia** (•left and below) (*the opposite of* **achlorophyllose** and **echlorophyllose**).

Leucobryum candidum leaf **chlorocysts**.
100 µm

Funaria hygrometrica **chlorophyllose** leaf.
100 µm

Hymenodon pilifer **chlorophyllose** spores.
100 µm

chloroplast — a small and usually round cellular organelle that contains chlorophyll. The cells of most mosses and liverworts contain many chloroplasts, whereas those of hornworts contain at most eight and usually only one or two (•below).

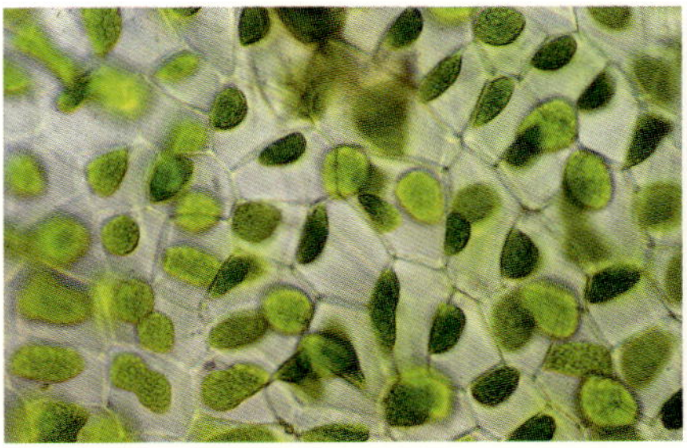

Anthoceros laminiferus thin slice of upper thallus surface, mostly one **chloroplast** per cell. 50 μm

Megaceros pellucidus thin slice of upper thallus surface, typically two **chloroplasts** per cell. 50 μm

chromatids — identical **chromosomes** produced by simple replication.

chromatogram — the final hard-copy printout of a chromatographic analysis.

chromatography — a technique for separating compounds by means of their different solubility in various solvents.

chromoplast — a cellular organelle that contains carotenoids or other **secondary pigments**.

chromosome — a structure inside the nucleus of a cell consisting of a very long molecule of double-stranded DNA and a roughly equal mass of protecting and supporting histone proteins. Chromosomes are the basic structural units of the genome in all creatures having nuclei that are bounded by membranes. The many segments along the prodigious length of chromosomal DNA include **genes** (which code for proteins) plus several kinds of non-coding portions. Those non-coding segments include operators and promoters, which control the transcription of the DNA into RNA, and introns, which interrupt genes into two or more segments that are called exons. Oddly, the introns are transcribed into RNA along with the exons, but they're snipped out by enzymes before the RNA is translated into protein. Ordinarily, chromosomes are visible only while they're replicating during **meiosis** or **mitosis**. During that brief period the DNA is "supercoiled", and as a result it's often packaged more tightly than relaxed interphase chromatin is by a "packing ratio" that's typically 1000–2000, and in some cases up to 7000.

cicatricate — scarred.

ciliate — having fine hairs (**cilia**) at a **margin** or on a surface (•below, left) (*see* **hairy** for a comparison of similar terms).

cilium (plural **cilia**) — a delicate hair- or tooth-like structure fringing a leaf, **thallus**, or other structure, or alternating with the **processes** of the **inner peristome** (**endostome**) (•below, middle).

cinderella — a common name given to the moss *Funaria hygrometrica* because it's one of the first to colonize the ash-laden soil of burned-over areas. It's also quick to appear on wet soils where there's been a camp fire. For those same reasons, in France it's called *la charbonniere*, which means *peddler of charcoal or coal.*

cinereous (or **cineraceous**) — colored ashy-gray.

circinate — strongly bent or curved into a circle or near-circle (•below, right).

Metzgeria sp. **ciliate** margins of thalli.
1 mm

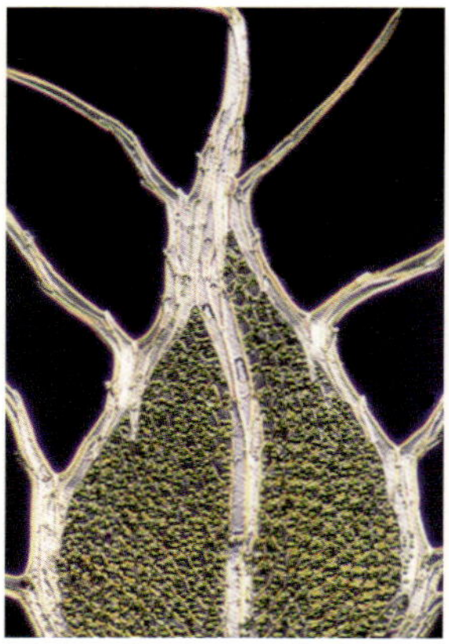

Catharomnion ciliatum leaf with marginal **cilia**.
100 µm

Entodon rubicundus **circinate** branches.
5 mm

circumboreal — a distribution pattern in which a **species** is widespread in but not necessarily restricted to the Northern Hemisphere (*synonym* circumpolar).

cirrate (or **cirrose** or **cirrhose**) — (1) in mosses, ending with a curled or wavy **appendage**, usually an extension of the **costa**, (2) curled or wavy.

cladautoicous — **monoicous** and with the male and female **inflorescences** on separate branches (for a comparison with other arrangements, *see* **autoicous**).

cladium (plural **cladia**) — a modified branch that acts like a **vegetative propagule**.

cladocarpous — in mosses and liverworts, said of an **archegonium** (and later the **sporophyte**) borne at the tip of a short specialized **lateral** branch.

clathrate — latticed (•below, left) (*synonym* **cancellate**).

clavate (or **claviform**) — club-shaped, elongate and thickened toward the tip (•below, middle and right).

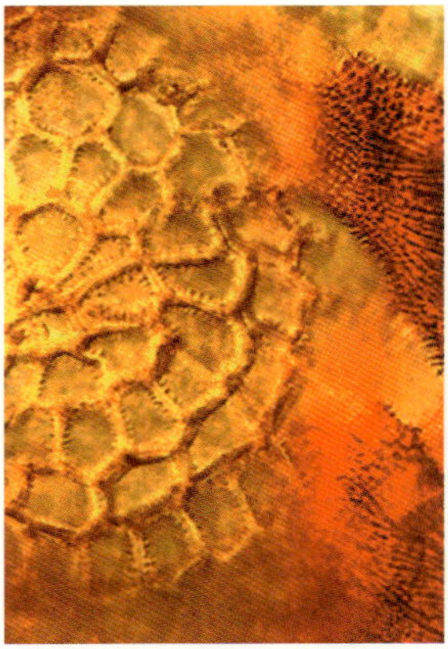

Funaria hygrometrica **clathrate** disk at tip of peristome teeth.
50 µm

Breutelia pendula **clavate** paraphyses.
50 µm

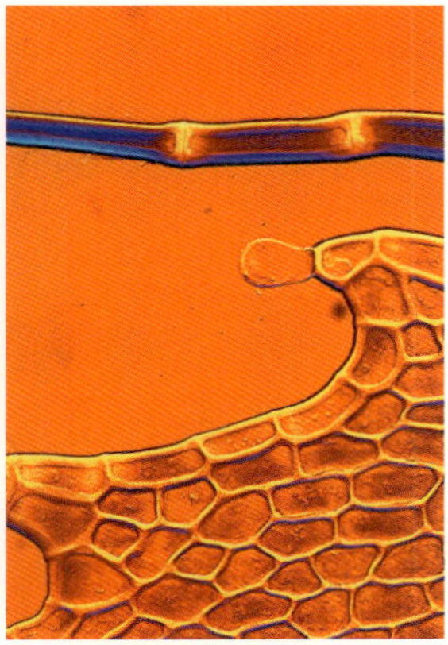

Schistochila attenuata **clavate** slime papilla.
100 µm

cleft — deeply divided, split, or incised (•below).

cleistocarpous — said of a moss **capsule** which has no **operculum** or **valves**, and therefore opens (**dehisces**) not along a line of built-in weakness but rather by the decay or breakdown of its wall (•left) (*similar to* **astomous**, lacking a mouth) (*the opposite of* **stegocarpous**, opening by means of a **lid**). Cleistocarpous dehiscence is also sometimes termed **irregular** (*the opposite of* **regular**).

cline — a graded range of genetic and phenotypic variation from one population to the next along a continuous geographic range.

clone — a population of genetically identical organisms produced from a single ancestor either naturally by asexual reproduction such as **apomixis** or artificially by propagation techniques like cuttings and grafts.

Pleurophascum grandiglobum var. *decurrens* **cleistocarpous** capsule.
1 cm

Lepicolea attenuata deeply **cleft** leaf.
100 µm

club-hair — a small **axillary filament** consisting of one long **terminal** cell atop a **basal** stalk of a few short cells (•below).

clump — a **tuft** of mostly **erect** plants as opposed to a **mat** of **prostrate** plants.

clypeate — shaped like a shield.

coacervate — heaped or piled together (*synonym* **cumulate**) (*compare with* **aggregate**, clustered together).

coalesced — fused together.

coarctate — pressed together, squeezed in, or **constricted**.

cochleariform — shaped like a spoon (•left) (*compare with* **cymbiform**, shaped like a shallow boat and therefore less deeply **concave**).

Lembophyllum divulsum **cochleariform** leaves. 1 mm

Meesia muelleri axillary **club-hairs** (wm). 100 µm

cochleate — twisted into a stout spiral like the shell of a snail.

coelocaule — in liverworts, a hollow **fleshy** structure that's derived from both stem and **archegonium** tissue and protects the embryonic **sporophyte**. It's a type of **perigynium**, and has the same function as a **perianth** and **bracts** (•below) (*synonyms* **stem-calyptra** and **shoot-calyptra**) (*compare with* **calyptra**, an embryo-protecting structure that's derived from only the wall of the archegonium).

Dendromastigophora flagellifera **coelocaule**. 1 mm

Paraschistochila conchophylla terminal **coelocaule**. 5 mm

cohering (or **coherent**) — sticking to an organ of the same kind (*the opposite of* **adhering**, sticking to an organ of a different kind) (*compare with* **connate**, fused with a structure or organ of the same kind, and **adnate**, fused with a structure or organ of a different kind).

collenchymatous — having cell walls thicker at the corners than on the sides (•below, left).

colliculate — covered with small, rounded elevations.

collum (plural **colla**) — in mosses, the **tapered** base of a **capsule** (•below, right).

Tetraphis pellucida **collenchymatous** median leaf cells. 100 µm

Leptostomum inclinans capsules with prominent **colla**. 1 mm

columella (plural **columellae**) — a central column of **sterile** tissue running up the central **axis** of a **sporangium** and surrounded by **spore**-producing tissue. It's developed particularly well in the hornworts, whereas in the *Sphagnum* mosses, it's reduced to merely a dome (•below, left and middle).

coma (plural **comae**) — a cluster of branches at the tip of a *Sphagnum* stem, or in other mosses a **tuft** of crowded leaves at the stem **apex** (•below, right).

Phaeoceros hirticalyx **columella** and spores. 50 µm

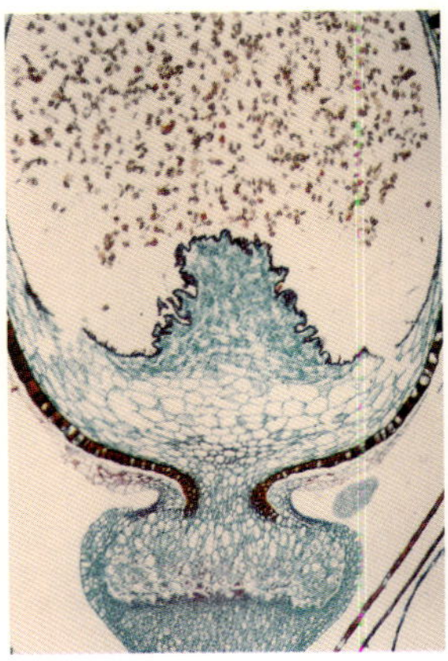

Sphagnum sp. **columella**. 100 µm

Hypnodendron comatum **coma** (top view). 1 cm

Ptychomnion aciculare **comite** collar in stem xs. 100 µm

comites — "stiffener" cells that support the leaves and stems of some mosses. They're long, slender, and thick-walled (fiber-like), and occur in bundles or cylinders in the **costa** and **borders** of leaves and toward the outside of stems (•left) (*synonym* **stereid**).

commensal — an organism that betters its growth and/or reproduction by living on or near another organism called the host. The host is neither harmed nor helped during the interaction (*see* **interactions**). Bryophytes that gain a place in the sun by perching on the stems or leaves of other plants are commensals.

commissural pores — in *Sphagnum*, pores arranged along the **margins** of the **hyaline** leaf cells (**leucocysts**).

commissure — in the leaves of species of *Sphagnum*, the **margins** of **hyaline** cells (**leucocysts**) where they adjoin the green cells. **Commissural pores** are arranged along the margins.

comose — (1) hairy (*see* **hairy** for a comparison of similar terms), (2) having **tufts** (**comae**) of large, crowded leaves at the stem **apex**. A stem which is highly comose can resemble a **rosette** (**rosula**).

compensation point — in plants, when the loss of energy from cellular respiration just equals the gain of energy from photosynthesis.

competitor — an organism or population that suffers a decline in its growth and/or reproduction by having to share scarce energy, materials, and space with others (which suffer similar losses) (•below) (*see* **interactions**).

Racomitrium lanuginosum, **competitor** with *Lindsaea linearis*, a fern. 1 cm

Leptotheca gaudichaudii, **competitor** with *Cladonia pleurota*, a lichen. 1 mm

Racomitrium ptychophyllum, **competitor** with *Raoulia* sp., a flowering plant. 1 cm

complanate — said of a leafy shoot that's flattened into one plane (•below).

Neckeropsis calcicola **complanate** shoot. 1 mm

Distichophyllum microcarpum **complanate** shoot. 1 mm

complicate — folded lengthwise (*compare with* **conduplicate**, folded lengthwise along the middle).
complicate-bilobed — folded lengthwise but also two-**lobed**, often with one lobe more complex (•next page and below, middle).
complicate-carinate — sharply folded lengthwise along a central keel.
compound pore (or **complex pore**) — a **pore** in some marchantialean liverworts that's composed of an internal barrel-shaped or **cylindric** structure plus several rings of **superficial** cells (•below, left) (*compare with* **simple pore**, which has no internal structure and is surrounded by only superficial cells).
compressed — flattened lengthwise, squeezed together, or condensed (•below, right).

Marchantia macropora **compound pores**. 1 mm

Porella elegantula **complicate-bilobed** leaves. 0.25 mm

Heteroscyphus multispinus **compressed** perianth. 1 mm

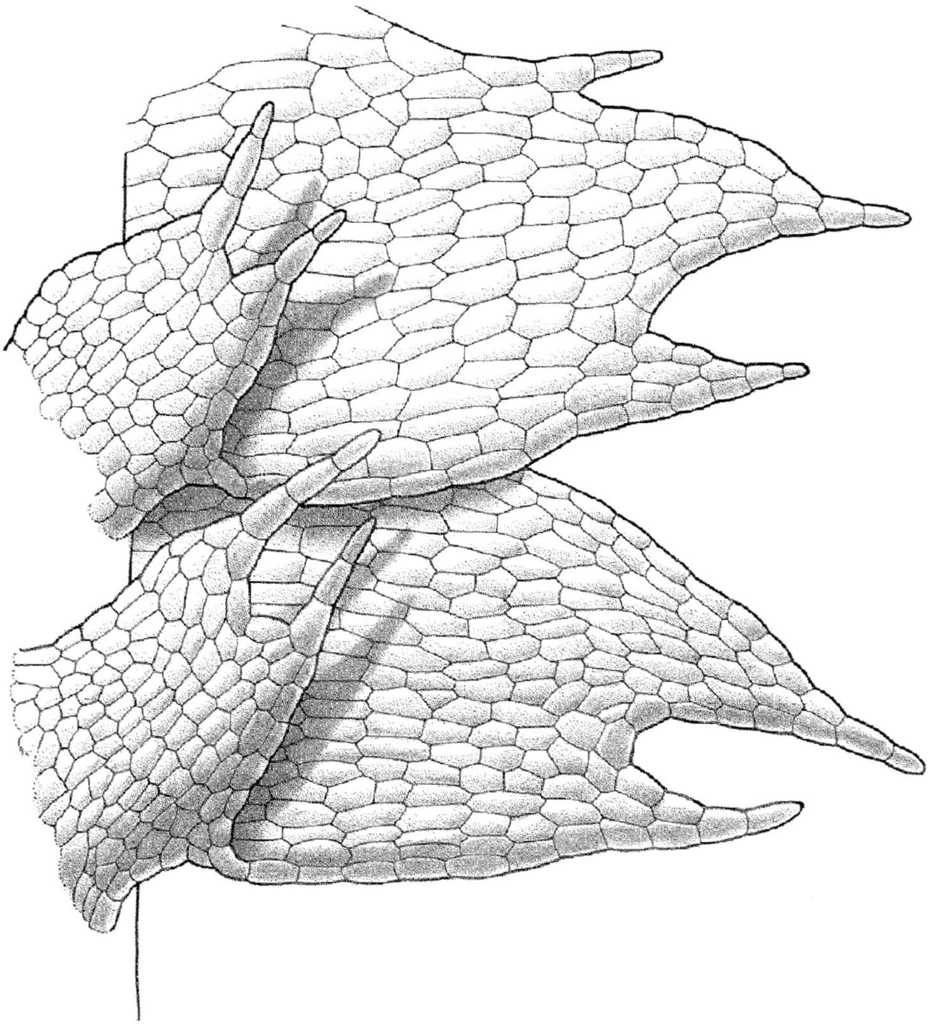

Balantiopsis diplophylla **complicate-bilobed** leaves. 100 μm

concave — curving inwards (•below, left and middle) (*the opposite of* **convex**, curving outwards).
concentric — having a common center.
conchiform — shaped like half of a bivalve mollusc (•below, right).

Neogrollea notabilis **concave** leaves.
1 mm

Acrolophozia pectinata **concave** trigones.
100 μm

Clasmatocolea sp. **conchiform** leaves.
1 mm

concolorous — said of two or more parts or structures that are colored the same.
conduplicate — folded lengthwise so sharply along the middle that the two sides are nearly parallel (*compare with* **complicate**, folded lengthwise).
confertate — crowded, dense, or pressed together closely (•below, left and middle).
confervoid — thread-like or loosely woven (•below, right).

Jungermannia inundata **confertate** growth habit.
5 mm

Bryum argenteum **confertate** growth habit.
5 mm

Lepidozia concinna **confervoid** habit.
1 mm

confluent — merging together seamlessly (•below).
congener — a member of the same **genus**.
congeneric — belonging to the same **genus**.

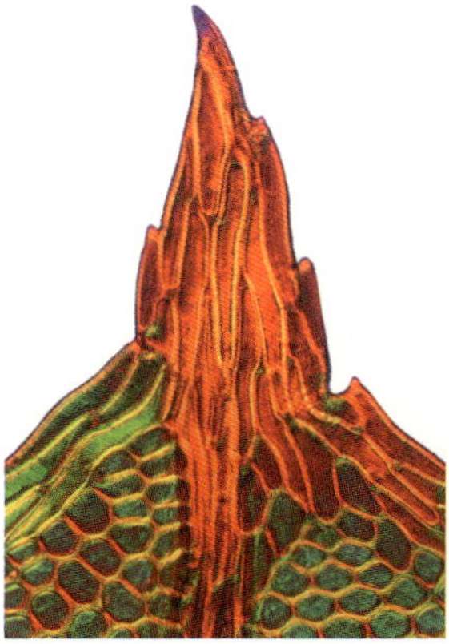

Plagiomnium cuspidatum costa tip **confluent** with border. 100 μm

Neogrollea notabilis **confluent** trigones. 100 μm

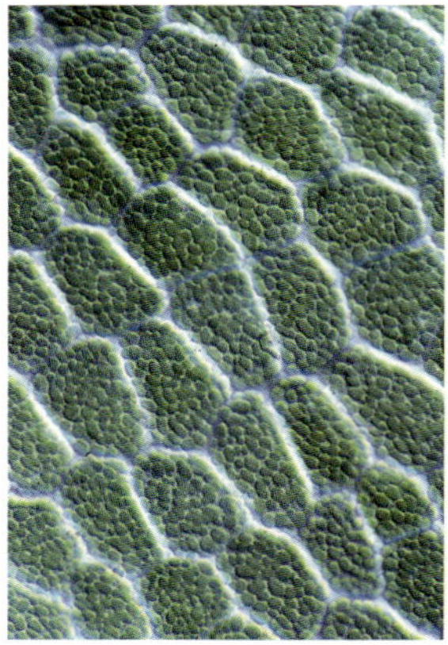

Plagiochila fruticella low **confluent** papillae. 50 μm

congested — crowded together (•below, left).
conglomerate — aggregated together.
conical — cone-shaped (•below, middle).
connate — fused with a structure or organ of the same kind (•below, right) (*the opposite of* **adnate**, fused with a structure or organ of a different kind) (*compare with* **cohering**, sticking to an organ of the same kind, and **adhering**, sticking to an organ of a different kind).

Spiridens veillardii **congested** leaves. 5 mm

Dawsonia superba **conical** operculum. 1 cm

Asterella australis **connate** pseudoperianth segments. 1 mm

connivent — meeting, converging, or approaching at the tips but not fusing (•below, left) (*compare with* **approximate**, near or close together but not touching or overlapping, and **contiguous**, touching or overlapping but not fused).
conoidal — resembling a cone.
conspecific — belonging to the same **species**.
constricted — **abruptly** or sharply narrowed or **contracted** (•below, right).

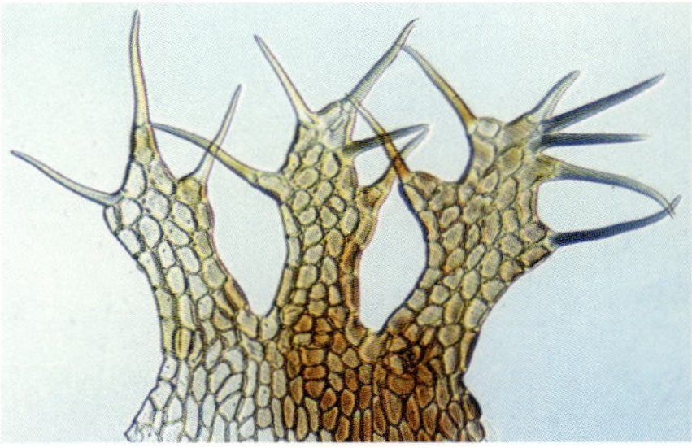

Herzogianthus vaginatus **connivent** leaf lobes. 100 μm

Ptychomnion aciculare **constricted** capsule (dry). 1 mm

contiguous — touching but not overlapping or fused (•below, left) (*compare with* **connivent**, converging at the tips but not fusing, and **approximate**, close together but not touching or overlapping).
contorted — (1) overlapping like shingles, (2) irregularly bent, twisted, and curved (•below, middle) (*similar to* **tortuose**).
contracted — **abruptly** narrowed or shortened.
convex — curving or bulging out (•below, right) (*the opposite of* **concave**, curving inwards).

Funaria hygrometrica **contiguous** peristome teeth. 50 μm

Bartramiopsis lescurii **contorted** leaves (dry). 1 mm

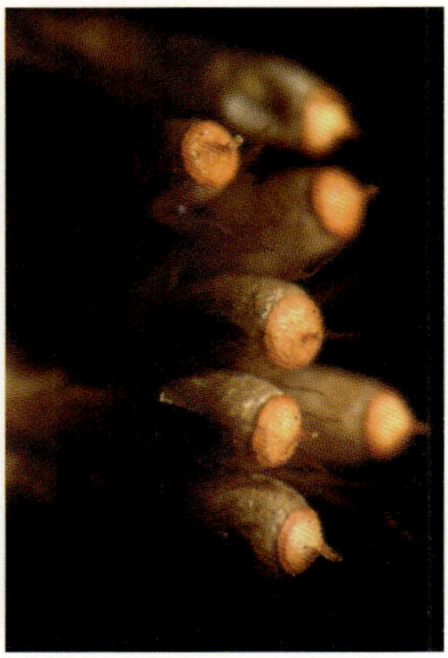

Venturiella sinensis **convex** capsule opercula. 1 mm

convolute — (1) rolled or folded to form a **sheath**, (2) rolled up inside a similar structure.

cordate — heart-shaped, said of leaves attached at the broad end (•below, left) (*similar to* **auriculate**) (*compare with* **obcordate**, heart-shaped but upside down, as in a leaf that's attached at its narrow end).

cord moss — a common name given to the **cosmopolitan** moss *Funaria hygrometrica* because its dried **seta** is so tightly twisted that it strongly resembles a cord.

coriaceous — leathery or tough.

cormophytes — any plants that produce **archegonia** (i.e. bryophytes, ferns and their allies, and the seed plants).

corneus — said of anything that's horny and hard but not so brittle that it can't be cut without breaking up or shattering.

cornute (or **corniculate**) — horned.

coronate — crowned.

corpuscle — an **axillary filament** consisting of one long **terminal** cell atop a **basal** stalk of two or three much shorter cells (*synonym* **club-hair**—*see* that entry for illustrations).

corrugate — regularly wavy or wrinkled (*compare with* **rugose**, strongly wavy, and **undulate**, somewhat wavy).

cortex — in a cross-section of a stem or branch of some bryophytes, (1) the outermost row or rows of cells, or (2) the tissue between the **central cylinder** and the **epidermis** (•below, middle and right).

Acrolejeunea allisonii **cordate** trigones.
100 µm

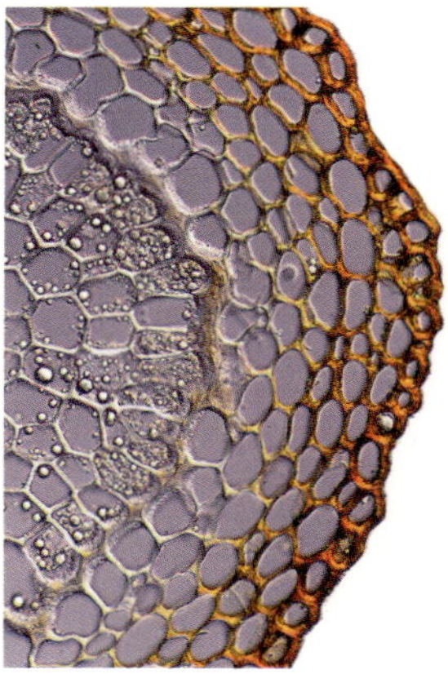

Dicnemon dixonianum stem xs showing **cortex**.
100 µm

Plagiochila sp. stem xs showing **cortex**.
100 µm

cortical — referring to or belonging to the **cortex**.

corticate — (1) having a **cortex**, (2) harder outside than inside.

corticolous — growing on bark (*synonym* **hylacolous**.

cosmopolitan — found in all major floristic zones of the world (•below). Many bryophytes can disperse long distances because (1) their **spores** are lightweight and small and so loft high into the atmosphere on rising currents of warm air, and many spores are resistant to long-term **desiccation** and ultraviolet damage, (2) they often produce **vegetative brood bodies** or other **propagules**, and (3) they're **totipotent** (even tiny fragments of their leaves and stems can sprout into new plants). Also, humans have vastly expanded the **habitats** of the many bryophytes that exploit disturbed sites—roadside cuttings, lawns, greenhouses, the cracks of sidewalks, and burned-over bogs and fields.

Funaria hygrometrica **cosmopolitan** moss.
5 mm

Polytrichum commune **cosmopolitan** moss.
1 mm

Ceratodon purpureus **cosmopolitan** moss.
5 mm

costa (plural **costae**) — in mosses and some liverworts, the thickened **midrib** or **nerve** of a leaf. The leaves of some mosses have a single or double costa or none at all (•below).

costate — having a **midrib** (**costa**) (*the opposite of* **acostate** and **ecostate**).

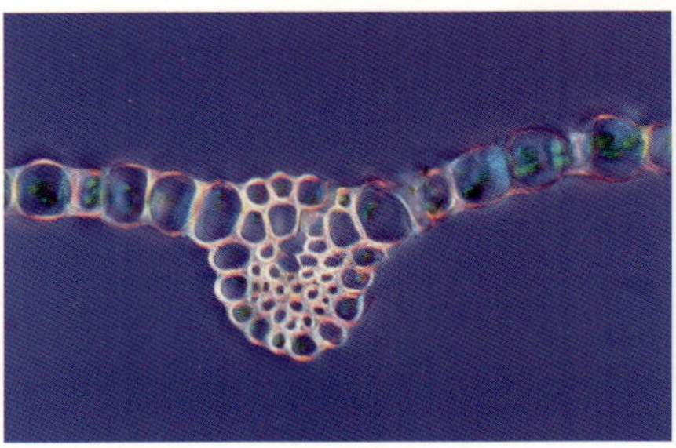

Bryum billardierei **costa** xs.
100 μm

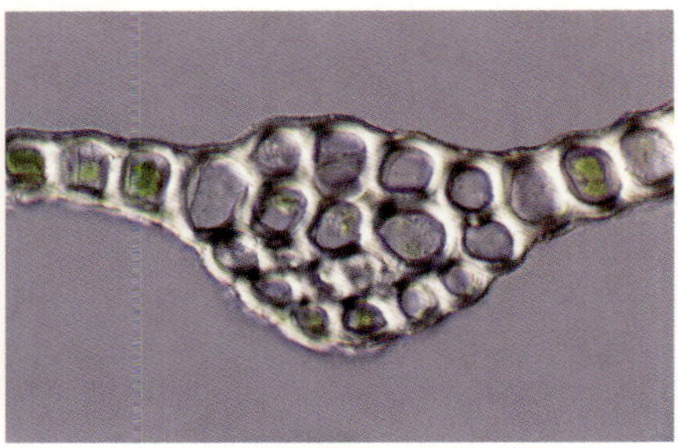

Racopilum convolutaceum **costa** xs.
50 μm

crassate — thick (•below) (*compare with* **incrassate**, thick-walled).
crateriform — shaped like a goblet.
creeping — growing along the **substratum** surface and anchoring at intervals.
crenate — having rounded **teeth** or scallops along the edge.

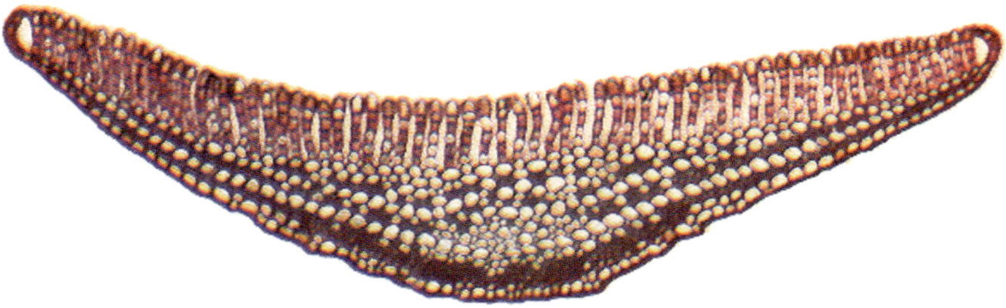

Dawsonia superba **crassate** leaf (xs). 100 µm

crenophilous — preferring a freshwater spring **habitat**.
crenulate — having minute rounded **teeth** or scallops along the edge, usually the bulging walls of individual cells (•below, left and middle).
cribbrose (or **cribrose**) — in mosses, (1) said of a **peristome** that resembles a sieve because it's finely perforated, (2) said of the triangular sieve-like area of large resorbed cells in the mid-base of stem leaves of some species of *Sphagnum*.
crisped (or **crispate**) — strongly curled, twisted, **contorted**, or wavy like crisp bacon (•below, right).

Polytrichum juniperinum **crenulate** lamella. 50 µm

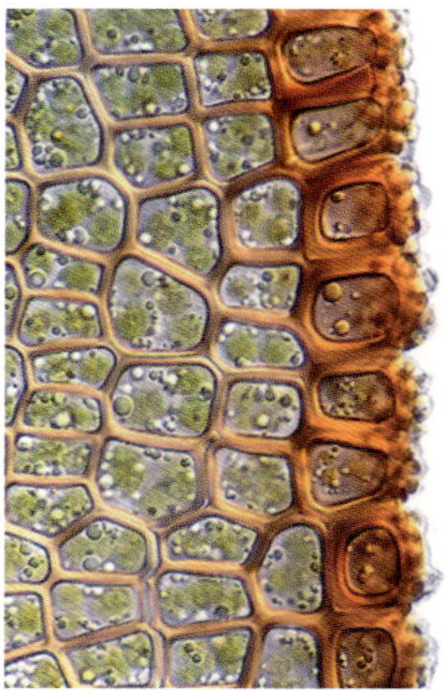

Psilopilum australe papillose **crenulate** lamella. 50 µm

Atrichum undulatum **crisped** leaves. 1 mm

cristate — (1) having a crest-like ridge resembling a rooster's comb (•below, left), (2) having a pale, **serrated margin** resembling a crest.
cruciate — in section, shaped like a cross.
crustaceus (or **crustaceous**) — resembling a crust, thin and brittle.
crymocolous — growing in tundra or polar regions.
cryptocoelocaule — a **coelocaule** with only a small hollow **fleshy** base and a short **perianth** at the top.
cryptopolar — said of a **spore** that germinates from the equator or either pole.
cryptopore — a **stomate** that's "hidden", deeply **immersed** in a depression or chamber formed by overarching adjacent cells (*the opposite of* **phaneropore**, a stomate that's not sunken).
C-shaped — said of **papillae** which appear crescent-shaped when viewed from above using an ordinary light microscope.
cubic (or **cubate**) — cube-shaped or nearly so.
cucullate — resembling or shaped like a hood (*synonym* **side-split**), (1) in leaves, with the **apex concave** and curved inwards to meet the sides, (2) in **calyptrae**, split up one side and conic, resembling a traditional monk's hood (•below, middle) (*compare with* **mitrate**, which is not split up one side and can be **entire** or **lobed** at its base).
cultiform — shaped like a plowshare.
cultriform — (1) shaped like a knife with parallel sides, (2) **asymmetrically** curved like a scimitar.
cum fructibus (abbreviated **c. fr.**) — Latin phrase meaning *with fruits*, hence producing **sporophytes** or sporulating (•below, right).

Calyptrochaeta cristata **cristate** seta.
1 mm

Leptostomum macrocarpon **cucullate** calyptra.
1 mm

Dicranum menziesii **cum fructibus**.
5 mm

cumulate — heaped or piled together (*synonym* **coacervate**).
cuneate — wedge-shaped (*synonym* **cuneiform**).

cuneiform — shaped like an upside-down triangle (*synonym* **cuneate**).

cupiform — shaped like a cask or tub (*compare with* **doliiform**, shaped like a barrel).

cupulate — cup-shaped with a mouth that's only slightly flared (•below, left) (*synonym* **cyathiform**) (*compare with* **campanulate**, bell-shaped with a mouth that's widely flared, and **urceolate**, pitcher-shaped, which is distinctly narrowed just below the mouth).

cupuliform — shaped like a cupola.

curvature — the bending or curving of a surface or structure (•below, middle and right).

Blindia lewinskyae **cupulate** capsule.
1 mm

Cryphaea tenella shoot **curvature**.
1 mm

Warnstorfia fluitans capsule **curvature**.
5 mm

cushion — a dense hemispherical or rounded **tuft** of crowded and radiating stems (•below). The cushion is a common growth form of bryophytes because it stores water, keeps out **competitors**, and resists flooding in streams or disturbance by grazing and pawing animals.

Bryum sp. **cushion** in alpine stream.
5 cm

Philonotis tenuis **cushion** (cut apart).
1 cm

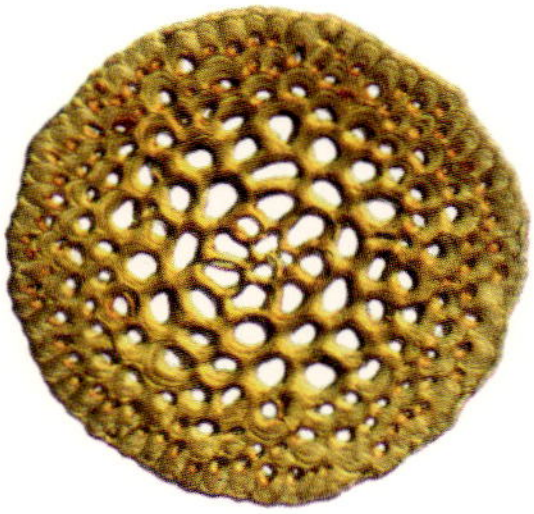

Campylopus clavatus **cutinized** seta (xs). 30 μm

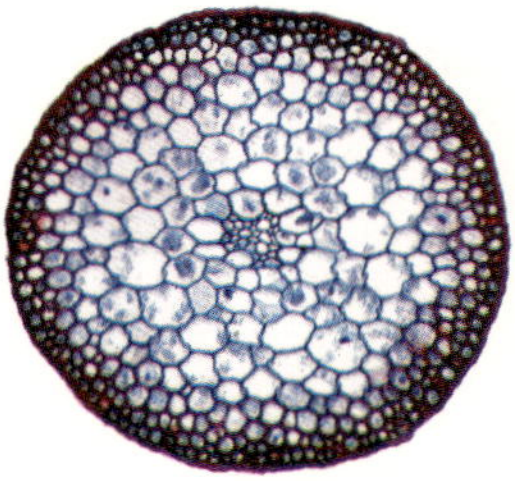

Polytrichum sp. **cutinized** seta (xs). 100 μm

Gymnomitrion cuspidatum punctate **cuticle**. 100 μm

cuspidate — (1) ending **abruptly** in a stout and rigid tooth-like cusp or point (•below, middle), (2) tapering gradually to a rigid point (•below, right).

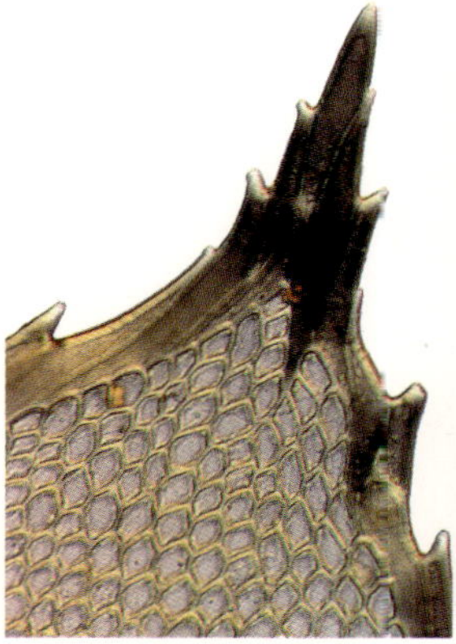

Hypopterygium japonicum **cuspidate** leaf apex. 50 μm

Calliergonella cuspidata **cuspidate** shoot tips. 1 mm

cuticle (adjective **cutinized**) — a non-cellular, waxy or fatty coating on the outermost wall of cells in direct contact with the environment, often roughened or ornamented (•left). In general, cuticle is only weakly developed in bryophytes. That trait, along with a large surface-to-volume ratio, boosts the uptake of water and dissolved mineral nutrients. In the moss family Polytrichaceae, however, the **adaxial** surface of the leaves and the uppermost cells of the **lamellae** are **cutinized**, so absorption is through the **abaxial** surface and the lower lamellae cells (•below).

cutin — a water-repellent fatty substance in **cuticle** (•top and middle left). The **spore** coat of mosses typically is cutinized. As well, the outermost cells of moss **setae** cutinize with age, steadily increasing the **sporophyte**'s dependence on the **gametophyte** for its water supplies. Although cutin in mosses functions primarily as waterproofing, it perhaps protects against too-strong sunlight as well—for example, the leaf in some **hair-cap mosses** (Polytrichaceae) curls in over the **lamellae**, reflecting some of the sunlight that would otherwise damage the cells of the lamellae or break down their chlorophyll.

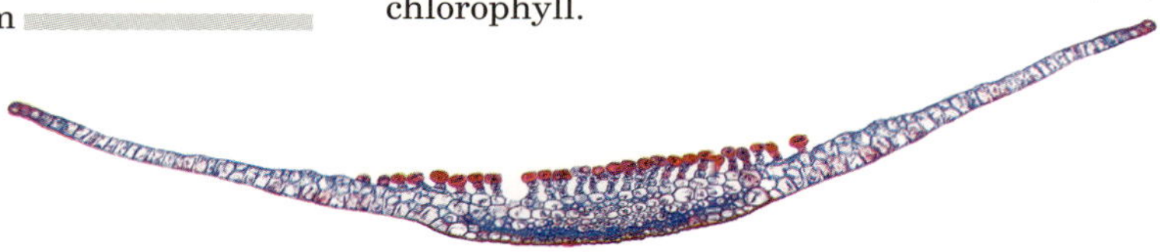

Polytrichum sp. partially **cutinized** leaf (xs). 100 μm

Bryum sp. shoots with spiraly arranged leaves.
1 cm

Fissidens pallidus shoot (UV microscope wm).
1 mm

cutting face — any surface of a cell where daughter cells are formed. For example, the **apical** cell of a leafy liverwort **shoot** is like an upside-down pyramid. Its three lower surfaces are cutting faces (*see* **tetrahedral apical cell**). The cells produced by those faces form three primordial segments called **merophytes**, which produce stem and leaf tissues.

The apical cell of a moss **gametophore** typically has three cutting faces. As a result, the stem cells (and eventually the leaves as well) are initially cut off in three **ranks** or rows at 120° around the stem **axis**. However, that strict three-ranked arrangement rarely shows up on a mature stem (•left). That's partly because the cutting faces divide in sequence rather than all at the same time, and so the stem **initials** are arranged in a spiral around the stem rather than in a straight line. As well, though, the daughter cells aren't all the same size, so the fourth stem initial doesn't line up directly below the first one. As a result, the leaves of most mosses appear to be spiraled around the stem in more than three rows.

Even in moss shoots that are strongly flattened into a single plane, the leaves are actually in three ranks, and they get their flattened look because their leaf bases twist on the stem as they develop. *Fissidens* is an exception—it has only two cutting faces, but even it begins life with three, losing one of them later (•left).

Because most mosses are built entirely by the plodding division of only a single apical cell with three cutting faces, they all end up looking pretty much alike. In contrast, the seed plants, ferns, and other vascular plants have been able to evolve a wide diversity of forms and structures because they grow from complex **multicellular meristems**.

cyathiform — cup-shaped with a mouth that's only slightly flared (*synonym* **cupulate**) (*compare with* **campanulate**, bell-shaped with a mouth that's widely flared, and **urceolate**, pitcher-shaped, distinctly narrowed just below the mouth).

cygneous — curved downwards at the top like the neck of a swan (•below).

Campylopus clavatus capsules with **cygneous** setae. 1 mm

cylinder — the **central strand** in the stem of some mosses and liverworts.

cylindric (or **cylindrical**) — shaped like a cylinder (•below).

Examples of **cylindric** capsules: (from left to right) *Bryum billardierei, Leptotheca gaudichaudii* (3), *Tetraphis pellucida, Tortula muralis,* and *Barbula unguiculata.*

cymbiform — shaped like a shallow boat (*synonym* **navicular**) (*compare with* **cochleariform**, shaped like a spoon, and therefore more deeply **concave**).

cytokinins — a class of plant hormones that trigger cell division. In their structure they resemble adenine, one of the four purine building blocks of DNA and RNA. **Kinetin** is the best-known cytokinin and is widely used in plant-growth experiments in the vascular plants as well as bryophytes, but it probably doesn't occur naturally. It's known to influence the growth of the **seta** of some mosses and also the formation of sex organs in a few mosses.

daedaleus (or **daedaleous**) — having a rough and stout **truncate** point.

day-neutral — said of a plant that doesn't respond to seasonal changes in the lengths of day and night.

dealbate — covered with an **opaque** white waxy or paint-like powder.

deciduous — in mosses and leafy liverworts, falling off, not persistent, or lost at maturity (mostly applies to the **annulus** or **operculum** of mosses, but sometimes to their leaves and **hair-points**, branches, or **brood bodies**) (*synonym* **caducous**).

declinate — bent or curved downwards or forwards.

decolorate — faded, discolored, turning pale with age or under harsh conditions.

decumbent — **prostrate** towards the base but **erect** or **ascending** towards the tip.

decurrent — (1) said of leaf **margins** which extend below the leaf's **insertion** as ridges or **wings** on the stem, (2) said of anything which extends below its point of origin on a structure (•left).

Racomitrium lanuginosum long-**decurrent** hair-point.
100 µm

decurved — curved downward.

dedifferentiated — said of a cell that has reverted from a mature, specialized state to an immature, unspecialized state.

deflexed — bent or curved downwards.

defoliate — **denuded** of leaves.

dehisce (noun **dehiscence**) — to open by the rupturing or splitting of **valves** (for example, the capsules of hornworts and most liverworts), or to open by the loss of an **operculum** (for example, the capsules of most mosses).

dehiscing (or **dehiscent**) — said of any structure that's rupturing or splitting open (•below).

deliquescent — said of a stem that's branched or divided so repeatedly or strongly that its tip can't be seen clearly.

Dendroceros validus horn **dehiscing** by two valves. 1 mm

Schistochila sp. sporangia **dehiscing** by four valves. 1 mm

Sphagnum sp. capsules **dehiscing** by opercula. 1 cm

deltoid — (1) shaped like an equilateral triangle, (2) shaped like any kind of triangle or a capital delta (Δ, the fourth letter in the Greek alphabet) (•below), (3) a solid that's triangular in cross-section.

Taxiphyllum alternans **deltoid** pseudoparaphyllium. 100 μm

dendroid — having a growth habit like a tree (•below, left and middle) (*synonym* **canopy former**).
dendrophilous — preferring a tree **habitat**.
dentate — with outward-facing **teeth** that can be **multicellular** (•below, right).

Hymenophyton flabellatum **dendroid** habit.
10 mm

Hypopterygium filiculaeforme **dendroid** habit.
5 cm

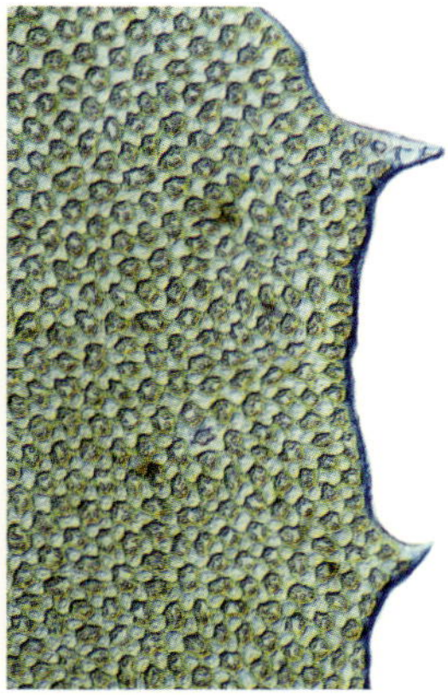

Plagiochila circinalis **dentate** leaf margin.
100 µm

denticulate — with fine **teeth**, often just cell tips projecting from the **margin** (•below, left).
denuded (or **denudate**) — with the leaves eroded or worn away (•below, middle).
deoperculate — said of a moss **capsule** after its **operculum** has fallen away (•below, right).
depauperate — stunted or poorly developed.
dependent — hanging down from its own weight.

Pendulothecium punctatum **denticulate** margin.
100 µm

Fontinalis antipyretica **denuded** stems.
0.5 mm

Bryum billardierei **deoperculate** capsule.
1 mm

Leptostomum inclinans leaf **deuters** (xs).
100 µm

depressed — (1) shallowly notched (*synonym* **retuse**) (*compare with* **emarginate**, which denotes a deeper notch), (2) flattened, low, or dwarfed.

derived — said of an organism that's presumed to have arisen from a more **generalized** ancestor.

dermatitis — inflammation of the skin. Some *Frullania* liverworts can cause severe dermatitis.

descending — sloping gradually downward.

desiccation — severe dehydration. Many bryophytes, especially mosses, can withstand desiccation, but only by suspending their growth and reproduction.

det. (abbreviation for **determinavit**) — Latin word meaning *he has identified it* but literally *enclosed within boundaries*, an abbreviation often noted on herbarium collections which have been identified by acknowledged experts.

deuters — in mosses, cells with large **lumina** stretching across the **nerve** (•upper left, and below) (*synonym* **guide cells**).

Campylopus clavatus leaf **deuters** (xs).
100 µm

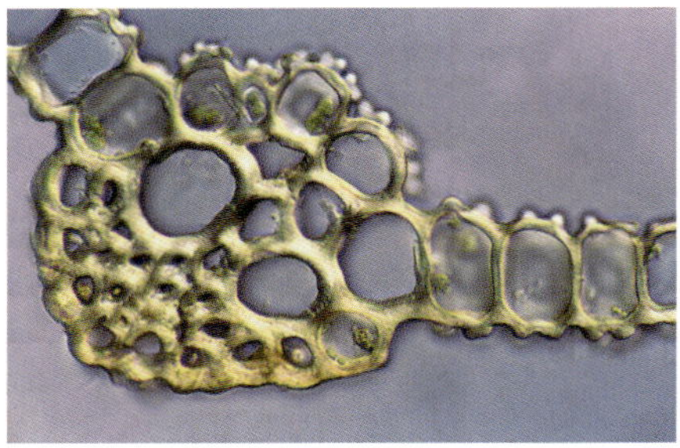

Tortula muralis leaf **deuters** (xs).
50 µm

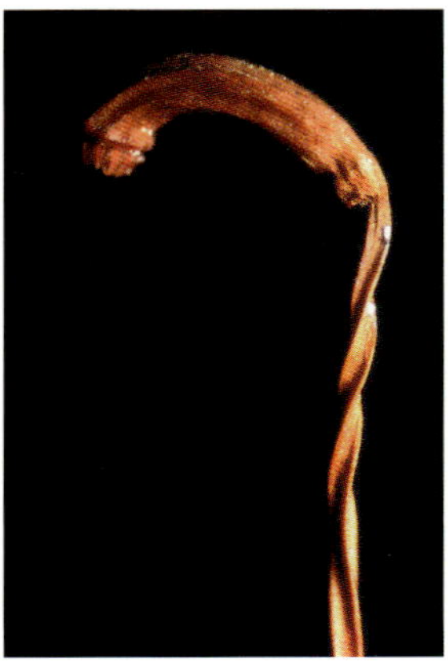

Racopilum convolutaceum **dextrorse** seta.
5 mm

dextrorse (or **dextrorsal**) — twisting upwards in a counterclockwise spiral (from left to right, as in a standard right-hand screw thread) (•left and below) (*the opposite of* **sinistrorse**, twisting upwards in a clockwise spiral from right to left, as in a nonstandard left-hand screw thread).

Standard **dextrorse**-threaded screws.

diandrous — bearing two **antheridia**.
diaphanous — fine and translucent.
diaspore — any structure by which a plant can reproduce itself (•below) (*synonym* **disseminule**). In bryophytes, diaspores can be sexually produced **spores**, asexually produced **brood bodies** such as **gemmae** and **tubers**, or just broken-off fragments of stems and leaves.

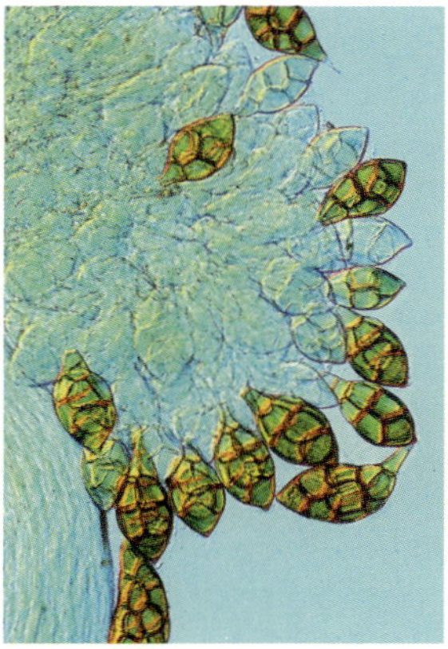

Aulacomnium androgynum **diaspores** (gemmae).
50 µm

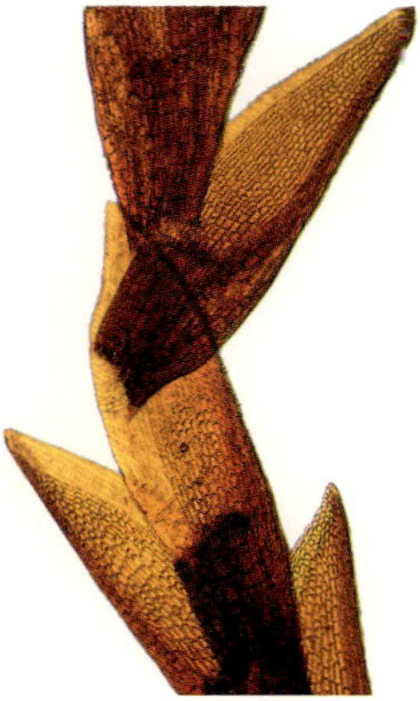

Dicranum flagellare **diaspore** (branch).
0.5 mm

Pohlia ochii **diaspore** (surculus).
100 µm

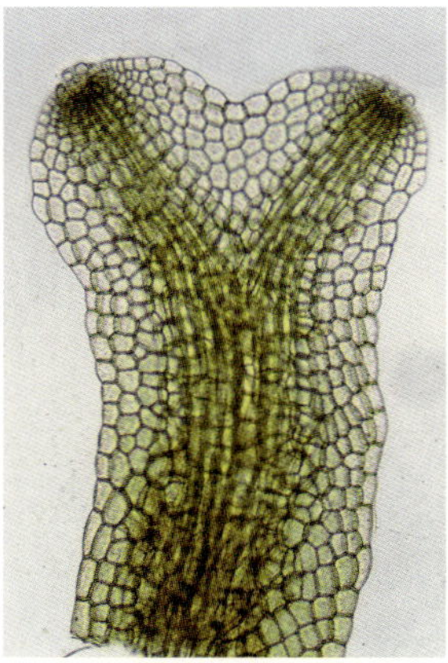

Metzgeria alpina **dichotomous** shoot branching.
0.5 mm

dichotomous — regularly dividing into two roughly equal branches or parts as the result of the division of an **apical** cell (•left) (*compare with* **pseudodichotomous**, branching which looks dichotomous but in fact is not caused by the division of an apical cell). True dichotomy is unknown in the mosses.
didymodontolith — a **tufa**-like mass deposited by the lime-loving moss *Didymodon tophaceus*. The ground and stream water in limy areas is saturated with calcium bicarbonate. The *Didymodon tophaceus* and other **calciphiles** remove carbon dioxide from the water during their photosynthesis. As a result, insoluble calcium carbonate precipitates out of the water and builds up on and around the plants as accretions called tufa. With time, the tufa hardens into **travertine** a porous brown limestone that in Roman times was used to build the Coliseum. In more modern times, it's been carved and polished into decorative staircases and floors for commercial and government buildings.
didymous — (1) divided into halves, (2) growing in pairs.
differentiated — (1) said of a cell that's mature and also specialized in its **morphology** and physiology, (2) said of cells that are distinctly different in size, shape, and/or color from nearby cells.

digammoid (or **digammiform**) — V-shaped (•below, left) (*compare with* **hyoid** or **ypsiloid** or **ypsiliform**, U-shaped or horseshoe-shaped).
digitate — having finger-like lobes radiating from the center (*synonym* **palmate**).
dimidiate — only half-formed.
dimorphic (or **dimorphous**) — having two different forms (•below, right).

Acrolophozia pectinata **digammoid** leaf sinus. 10 μm

Kindbergia praelonga **dimorphic** lanceolate stem leaf (left) and cordate lateral leaf (right). 0.5 mm

dioicous — producing male and female sex organs (**antheridia** and **archegonia**) on separate plants (•below) (*the opposite of* **monoicous**) (*not the same as* **dioecious**, which applies to the **diploid sporophytes** of vascular plants).

Polytrichadelphus magellanicus, **dioicous** (male). 1 cm

Haplomitrium gibbsiae, **dioicous** (male & female). 1 mm

Leptostomum macrocarpon, **dioicous** (female). 1 mm

diploid — said of a cell which contains two complete sets of **chromosomes** (denoted by 2n) in the nuclei of its cells—typically, bryophyte **sporophytes** are diploid (*compare with* **haploid**, said of a cell which contains only one complete set of chromosomes (denoted by n or 1n) in the nuclei of its cells—usually, bryophyte **gametophytes** are haploid) (*see* **ploidy level**, **polyploid**, and **polyploidy**).

Psilopilum crispulum **discoid** male "flowers" filled with sperm-bearing fluid. 5 mm

Breutelia pendula **discoid** male "flowers" (dry). 1 cm

diplolepidous — (1) said of a **peristome** with two distinct rings of **teeth**, an inner **endostome** and an outer, more heavily thickened **exostome** (*synonym* **double peristome**) (*the opposite of* **haplolepidous**), (2) said of a jointed (**articulated**) peristome tooth in which each segment of its outer face is formed from remnants of the walls of two cells.

diplospory — (1) the formation of **spores** that are **diploid** (2n or "unreduced"), (2) **polyploidy** arising from diploid spores.

discoid (or **disciform**) — flattened into a plate or disk (•above, and below, right).

discus (or **disc** or **disk**) — the broad base of a leaf below any **lobes**, akin to the palm of a hand below the fingers (•below left).

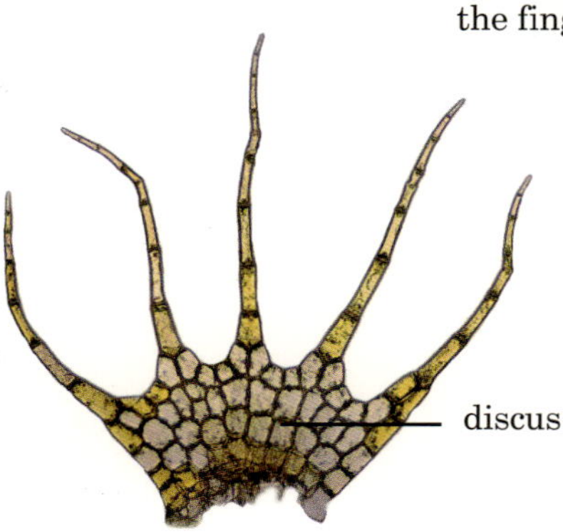

Telaranea gibbsiana leaf **discus**.
100 μm

Tetraphis pellucida **discoid** gemmae.
100 μm

disjunct — a distribution pattern in which a **species** is found in widely separated areas.
dispositio — a way of describing the arrangement of leaves on a stem (*synonym* **phyllotaxy**).
dissected — (1) notched at the tip, (2) divided into many slender segments.
disseminule — any structure by which a plant can reproduce itself. Moss disseminules are highly diverse, ranging from sexually produced **spores**, asexually produced **brood bodies** such as **gemmae** and **tubers** (• left and below), to just broken-off bits of leaves and stems (*synonym* **diaspore**).

Pohlia ochii **disseminules** (fallen surculi).
1 mm

Bryum dichotomum **disseminule** (brood body).
50 µm

Metzgeria disciformis **disseminule** (gemma).
100 µm

distad — in the direction of further away (*the opposite of* **proximad**, closer).
distal — located at the top of a structure or the end that's furthest from its base or origin (its point of attachment), for example in a typical **spore**, the side which faced outwards in the original **tetrad** of spores and which therefore lacks a **triradiate ridge** (*the opposite of* **proximal** and **basal**).
distant — spaced well apart (*synonym* **remote**).
distichous — arranged in two rows or **ranks** on opposite sides of a stem (• below) (*synonym* **bifarious**).

Rhizogonium distichum **distichous** shoots.
1 mm

Pendulothecium punctatum **distichous** shoot.
3 mm

Chiloscyphus echinellus **distichous** shoots.
1 mm

Bryum sp. **divisural line**. 50 µm

divaricate — attaching at almost a 90° angle.

divergent — (1) spreading from a point of attachment (origin), (2) spreading in opposite directions.

divisural line — in mosses, the midline of a **peristome tooth**, usually zig-zag but sometimes **furrowed** (•left).

dolabriform — shaped like an ax or hatchet, with one end thick and straight-sided but the other end thin and double-**convex**.

doliiform — shaped like a barrel (*compare with* **cupiform**, shaped like a cask or tub).

dorsal — said of the upper surface of a **prostrate** stem, the outer surface of a **peristome tooth**, the lower surface of a leaf, and the upper surface of a **thallose** liverwort (•below, left) (*the opposite of* **ventral**).

dorsal lamina — in a *Fissidens* leaf, the **wing** opposite the **sheathing** base or **vaginant lamina** (*see* **ventral lamina** for illustration).

dorsal plate — the remnant wall thickening on the outer surface of an **exostome** tooth.

dorsiventral — flattened top to bottom, producing an upper (**dorsal**) and lower (**ventral**) surface (•below, middle and right).

Marchantia foliacea **dorsal** thallus surface. 1 mm

Heteroscyphus cymbaliferus **dorsiventral** shoots. 1 mm

Paraschistochila conchophylla **dorsiventral** shoots. 5 mm

double peristome — in some mosses, a **peristome** composed of two separate rings of **teeth**, with 16 **lanceolate** teeth, cross-striolate at the base and **trabeculate** at the back, and a well-developed **endostome** consisting of a high **basal membrane** and 16 lanceolate, **keeled segments** alternating with one or more **cilia** (•next page) (*synonyms* **hypnoid peristome** and **diplolepidous peristome**) (*the opposite of* **single peristome** and **haplolepidous peristome**).

double peristome (continued)

Cryptopodium bartramioides **double peristome.** 1 mm

Hypnodendron comatum **double peristome.** 1 mm

Warnstorfia fluitans **double peristome.** 1 mm

doubly serrate — having **teeth** in two adjacent rows, overlapping and joined in pairs.

doubly toothed — having **teeth** which are themselves toothed (*not the same as* **bidentate**, which means having two teeth, nor **geminate**, meaning having pairs of teeth).

dung mosses — mosses which grow on dung, all of them members of the moss family Splachnaceae (the genera *Splachnum*, *Aplodon, Tetraplodon*, and *Tayloria*). Dung mosses have several unique **adaptations** that ensure their survival—(1) their **capsules** release volatile organic acids that smell like dung or carrion, attracting flies and other insects, (2) their **spores** are small and clumped and readily stick to insect hairs, and (3) their **protonemata** and shoots tolerate osmotically concentrated dung liquids which kill competing mosses.

duplicate — arranged or growing in pairs.

dwarf male — in mosses, a miniature male **gametophyte** borne on a full-sized female plant (•left).

echinate — roughened by spiny projections, bristly hairs, or prickles (•below) (*see* **hairy** for a comparison of similar terms).

Ptychomnion aciculare **dwarf male** plant. 0.5 mm

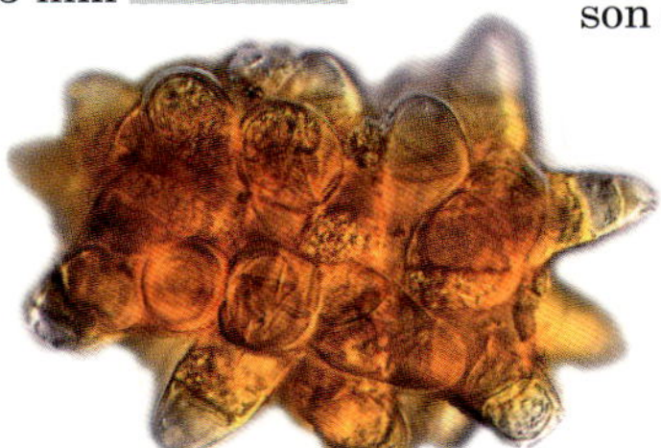

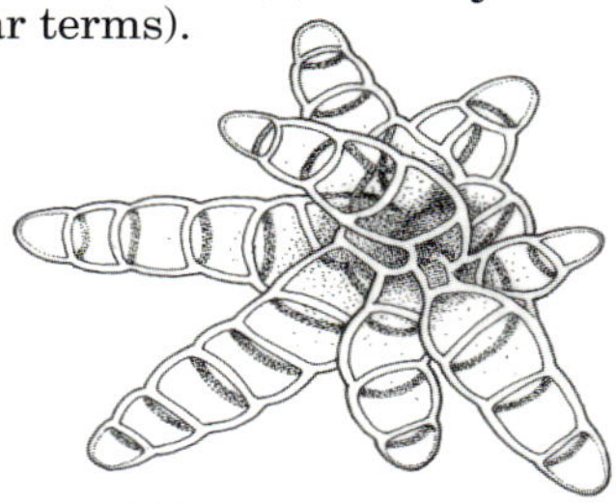

Hyophila involuta **echinate** gemma. 100 μm

echinulate — roughened by minute spines.
echlorophyllose — lacking chlorophyll (•below, left) (*synonym* **achlorophyllose**) (*the opposite of* **chlorophyllose**, containing chlorophyll).
ecostate — in mosses, lacking a **costa** (*synonym* **acostate**) (*the opposite of* **costate**).
ectohydric (or **exohydric**) — conducting water externally (•below, middle and right) (*compare with* **endohydric**, conducting water internally, and **mixohydric**, conducting it both internally and externally).

Lunularia cruciata gemma margin **echlorophyllose**. 50 µm

Ptychomnion aciculare **ectohydric** water. 5 mm

Cyathophorum bulbosum **ectohydric** water. 5 mm

edaphic — pertaining to soil.
edentate — lacking **teeth**.
egg — the **haploid gamete** (**ovum**) produced by the female sex organ (**archegonium**) (•left).
elate — unusually tall.
elater — a long, spirally thickened cell which changes shape markedly with **humidity** (•below).

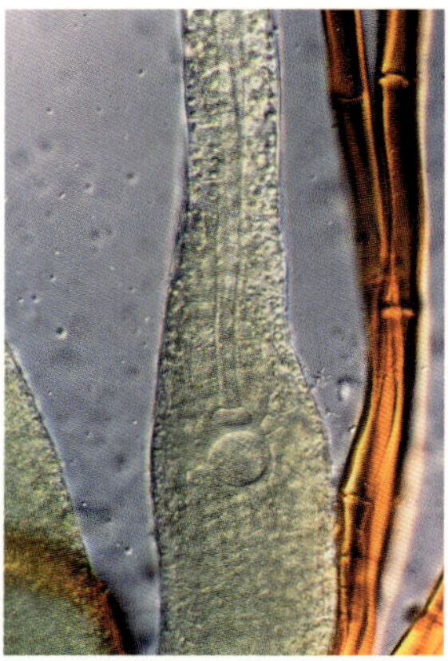

Bryum billardierei **egg** at base of archegonium. 100 µm

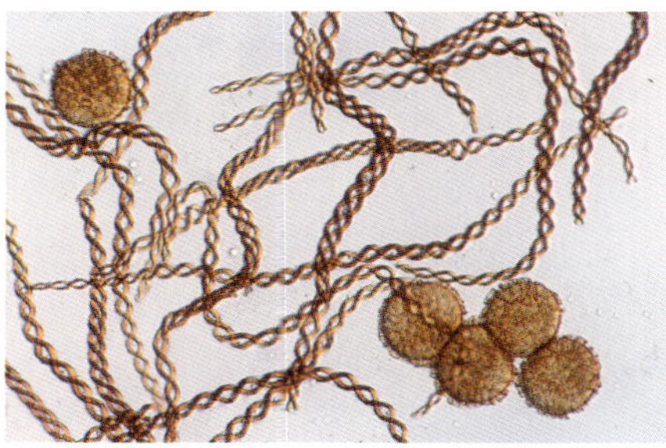

Fossombronia sp. **elaters** and spores. 100 µm

elaterophore — in the **base** or **apex** of the **capsules** of some metzgerialean liverworts, **sterile** tissue where **elaters** are attached.

elf cap moss — a common name given to species of *Buxbaumia* because of the fancied resemblance of their mature **capsules** to the "elf caps" of ancient fairy tales.

elimbate — lacking a **border** (•below, left).

ellipsoid — a solid that has an **elliptical** outline.

ellipsoidal — solid with an **elliptical** outline (•below, middle).

elliptic (or **elliptical**) — shaped like an ellipse, with a wide middle and rounded ends (•below, right).

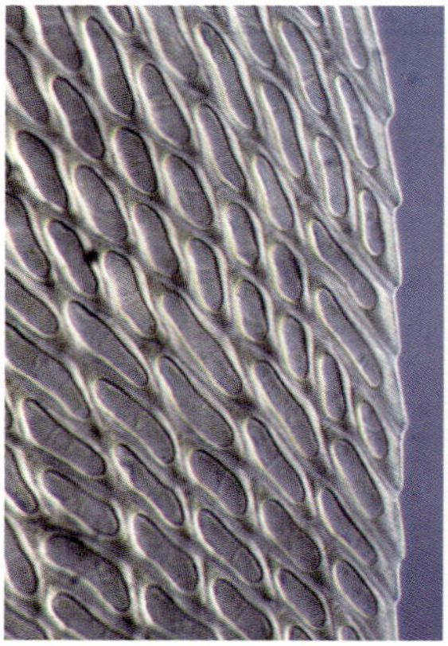

Lembophyllum divulsum **elimbate** leaf margin. 50 µm

Orthotrichum tasmanicum **ellipsoidal** capsule. 1 mm

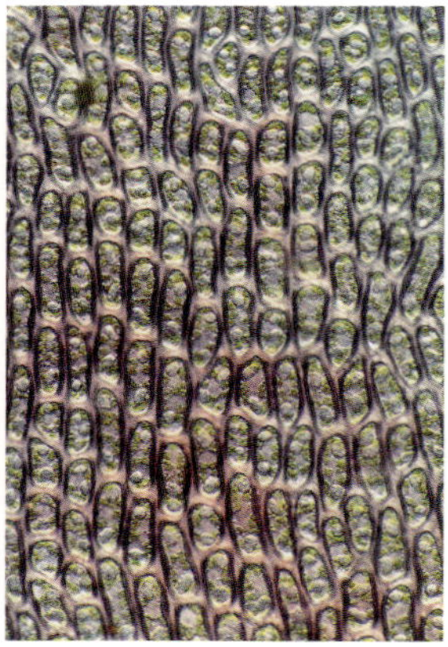

Clandarium xiphophyllum **elliptic** leaf cells. 100 µm

emarginate — broadly notched at the **apex** (*compare with* **retuse**, more narrowly notched).

embedded — fixed firmly and sunken deeply in a surrounding solid mass.

embryo — the immature and undeveloped stage of a **sporophyte**.

embryophyte — the name given to any plant that bears an **embryo** during its life cycle. The embryophytes include a diverse group of plants: liverworts, mosses, hornworts, ferns, horsetails, lycopods, gymnosperms, and angiosperms. They all share two traits: (1) their life cycle alternates between two **multicellular** generations, one of them called a **gametophyte** because it produces **gametes** (by **mitosis**) and the other called a **sporophyte** because it produces **spores** (by **meiosis**), and (2) they enclose their gamete-producing cells inside a **sterile jacket**. The **egg**-containing jacket not only protects the egg, but also shuttles nutrients to the developing **embryo** and goes on to influence the growth of the juvenile sporophyte.

emergent — partly exposed, hence said of (1) moss **capsules** that protrude only partly beyond the tips of the **perichaetial** leaves, or liverwort capsules that emerge beyond the tips of **perianths**, **bracts**, **bracteoles**, or other **sheathing** structures (• below, left) (*compare with* **immersed**, hidden within surrounding structures, and **exserted**, lifted well beyond them), (2) moss **cushions** along lake edges.

endemic — a distribution pattern in which a **species** or other **taxon** is restricted to a single country or **floristic** area.

endemism — the restriction of a **species** or other **taxon** to a single country or **floristic** area.

Venturiella sinensis mature **emergent** capsules. 1 mm

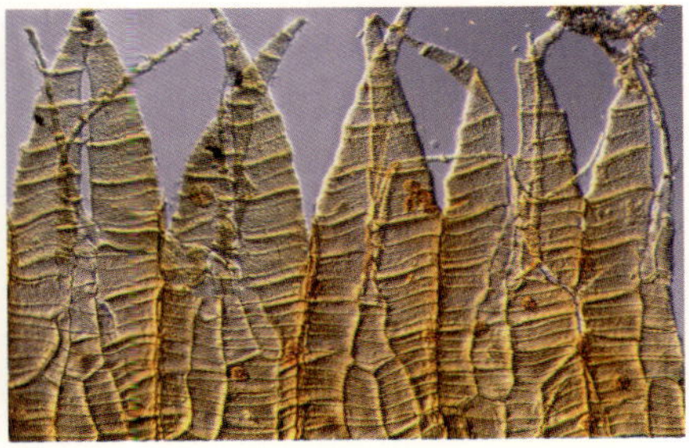

Racopilum aristatum **endostome** and basal membrane. 50 µm

endogenous — originating from inner cells rather than from surface cells (*the opposite of* **exogenous**, originating from surface cells).

endohyalocyst — (1) a swollen, empty, **hyaline** cell in the leaf base of species of the moss family Calymperaceae (*synonym* **cancellina**), (2) a **hyalocyst**.

endohydric — conducting water internally (*compare with* **ectohydric**, conducting water externally, and **mixohydric**, conducting it both internally and externally).

endosporic — said of a **spore** that divides several times before its coat ruptures (*compare with* **exosporic**, said of a spore that divides only *after* its coat ruptures).

endostome — in many mosses, the **inner peristome**, usually arising from a **basal membrane** and consisting of **segments** alternating with **cilia** (• above, right).

endothecium — in an **embryonic sporangium**, the cells inside the **amphithecium** that give rise to tissues other than the sporangium **jacket**, such as the **spore sac** and the **columella**.

enervate — in mosses, lacking a **midrib** (**costa**) (*synonyms* **ecostate** and **nullinervate**).

ensiform (or **ensate**) — shaped like a sword, straight with an **acute** point.

entally — toward the inside or interior.

entire — said of a **margin** that's smooth, lacking any **teeth**, **cilia**, indentations, fringes, or erosions (• below).

Schwetschkeopsis robustula **entire** leaf margin.
100 µm

Tetraphis pellucida **entire** leaf margin.
100 µm

Schistidium apocarpum **entire** leaf margin.
100 µm

ephemeral — (1) short-lived—as a group, the bryophytes have small and ephemeral **sporophytes**, (2) an organism which completes its life cycle within a single season (*similar to* **ruderal**). Leaf-dwelling (**epiphyllic** or **foliicolous**) mosses, lichens, and liverworts are ephemeral, as are **pioneer mosses** that invade freshly disturbed sites like burns, roadside clay banks, and patios with sand grouting separating the pavers. They're often displaced in the second season by more aggressive plants (• below).

Acaulon integrifolium, an **ephemeral** moss.
5 mm

Tortula truncata, an **ephemeral** moss in the cracks of a brick footpath.
10 cm

epicoelocaule — a **coelocaule** which has unfertilized **archegonia** at its top and is covered with **paraphyllia** and/or **scales**.

Polytrichadelphus magellanicus **epiphragm**.
0.5 mm

epidermis — the outermost layer of cells at the surface of a tissue (*synonym* **exothecium**).

epigaeus (or **epigeous**) — (1) growing close to the ground, (2) growing on soil.

epigonium — in mosses, the protective envelope of a developing **sporophyte** before its rupture into two parts, a **basal sheath** (**vaginula**) and upper **calyptra**.

epilithic — growing on rock (*synonyms* **epipetreous**, **petrocolous**, **rupestral**, and **saxicolous**).

epipetreous — growing on rock (*synonyms* **epilithic**, **petrocolous**, **rupestral**, and **saxicolous**).

epiphragm (plural **epiphragmata**) — in the **haircap mosses** (family Polytrichaceae), a circular membrane formed by the expanded tip of the **columella** and attached to the ends of the **peristome teeth**. It partially closes the mouth of the **capsule** after the **operculum** has dropped away (•left).

Austrolejeunea olgae, **epiphyll**.
1 mm

Siphonolejeunea nudipes, **epiphyll**.
1 mm

epiphyll — a plant which grows on the living leaves of another plant. Most epiphylls are liverworts or lichens, but many mosses colonize leaves at least occasionally, among them species of *Calymperes, Crossomitrium, Ephemeropsis, Leucophanes, Mitthyridium,* and *Syrrhopodon*. Epiphyllic liverworts are most common in the tropics, and many belong to the Lejeuneaceae. Epiphylls are as few as they are because leaves typically are short-lived, and many have smooth or waxy surfaces that mosses and liverworts can't readily **adhere** to. Successful epiphylls typically are small and **prostrate**, have **rhizoids** with adhesive disks, mature quickly (*see* **ephemeral** and **ruderal**), and reproduce by **gemmae** or other **propagules**. Some colonize only one particular part of leaves such as the **midrib** or the **margin** (•above) (*synonym* **foliicole**).

epiphyllous (or **epiphyllic**) — growing on the living leaves of another plant (*synonym* **foliicolous**) (*see* **epiphyll**).

epiphytic — growing on another plant but not parasitizing it (•below).

epithet — a **species** name, the second name of a Latin binomial (a two-word scientific name), for example *major* of *Psiloclada major*.

Colura saccophylla **epiphytic** on the moss *Dendroligotrichum dendroides*. 1 mm

Psoroma asperella, a lichen **epiphytic** on the moss *Breutelia pendula*. 1 cm

equidistant — uniformly or regularly spaced.

equilateral — said of any structure with sides that are nearly equal in length.

equitant — said of a leaf that sheaths a stem closely and/or tightly.

erect — (1) vertical or nearly so (•below, left and middle, and next page), (2) curved upwards (*see* **spreading** for a comparison of similar terms).

erecto-patent — making an angle with the stem of (1) less than 25°, (2) 20–45° (•below, right) (*see* **spreading** for a comparison of similar terms).

erect-spreading — making an angle of about 45° with the stem (*see* **spreading** for a comparison of similar terms).

Ceratodon purpureus **erect** shoots and setae. 1 cm

Clasmatocolea sp. **erect** shoots. 1 cm

Cryptopodium bartramioides **erecto-patent** leaves. 5 mm

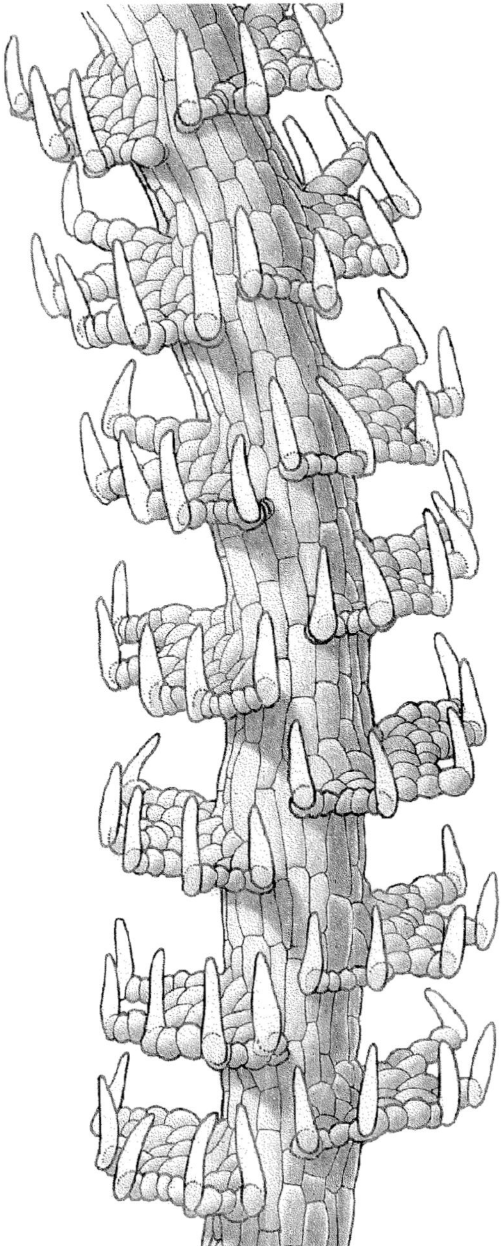

Psiloclada clandestina, **erect** terminal cells of leaf lobes. 100 µm

eremophilous — thriving in a desert **habitat**.
eroded — worn away.
erose — ragged or irregularly notched as if gnawed by an animal (•below, left).
-escens — a suffix meaning *becoming* or *turning*, as in **rufescent**, meaning *turning red*.
etiolated — said of abnormally elongated and pale leaves or stems, usually a result of low light.
euchromatin — region of a **chromosome** that's not condensed between divisions.
eutrophic — rich in mineral nutrients, particularly nitrate and phosphate.
everted — turned inside out.
exannulate — lacking an **annulus**.
exasperate — roughened (*synonyms* **scabrid**, **scabrous**, and **scabrose**).
excavate — **abruptly concave** or hollowed out.
excurrent — (1) in mosses, said of a **costa** that extends beyond the **apex** of the leaf (•below, middle) (2) said of anything that's clearly in the center of an organ.

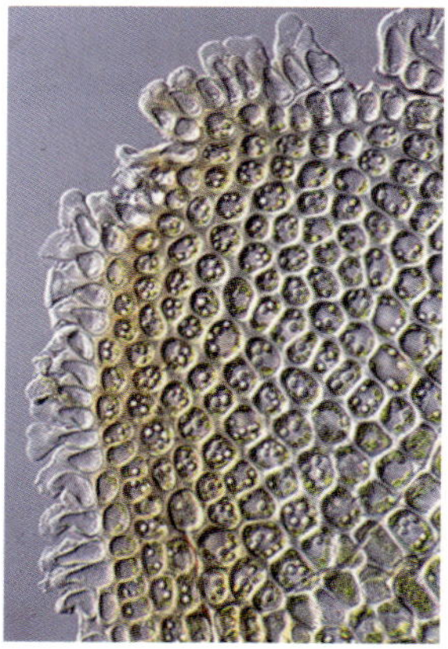

Nothogymnomitrion erosum **erose** leaf margin. 100 μm

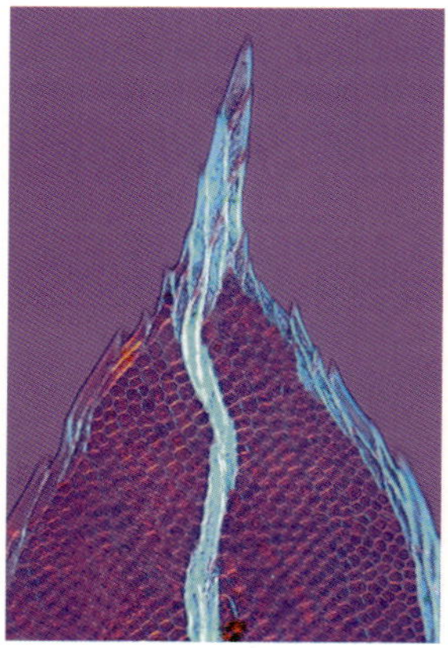

Lopidium concinnum **excurrent** leaf costa. 100 μm

Campylopus introflexus **exostome** teeth. 100 μm

exogenous — originating from surface cells (*the opposite of* **endogenous**, originating internally).
exohydric — (*see* **ectohydric**).
exospore — the outer layer of a **spore** wall, often ornamented.
exosporic — said of a **spore** that divides only after its coat has been ruptured (*compare with* **endosporic**, in which a spore undergoes several cell divisions before its coat ruptures).
exostome — the outer row or rows of a **double peristome**, consisting of **teeth** that usually are **forked** toward the top (•above, right) (*compare with* **endostome**, the **inner peristome**, consisting of **segments**).

exothecium (plural **exothecia**) — the outermost layer of cells in the **sporangium**) (•below).

Dicnemon dixonianum **exothecium** wall (xs). 50 μm

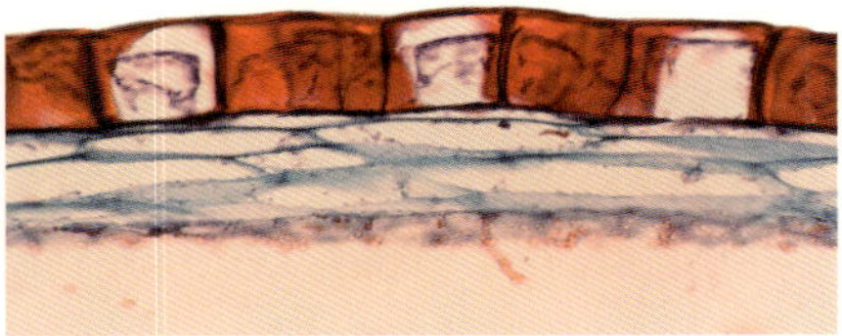

Sphagnum sp. **exothecium** wall (xs). 50 μm

explanate — flattened (•below, left) (*the opposite of* **inflated** or swollen).

exserted — protruding well beyond the tips of surrounding leaves, **bracts**, or other structures, and as a result completely exposed (•below, middle and right) (*compare with* **immersed**, hidden among surrounding structures, and **emergent**, protruding partly beyond them).

exsiccata (plural **exsiccatae**) — a set of dried and labeled specimens distributed to herbaria or collectors as standards for research.

exsiccatum (plural **exsiccata**) — a single dried specimen.

Frullania explanata **explanate** leaf lobule. 0.5 mm

Bryum argenteum mature **exserted** capsules. 1 mm

Fossombronia reticulata **exserted** sporophyte. 1 mm

extinguisher moss — a common name given to the moss *Encalypta* because its **calyptra** looks like a candle-snuffer.

extrorse — turned or bent outwards or away from the center (*the opposite of* **introrse**) (*compare with* **antrorse**, facing forwards or towards the tip, and **retrorse**, facing backwards or away from the tip).

facies — general appearance (*synonym* **habit**).

facultative — not restricted to one host or **substratum** (*the opposite of* **obligate**).

failing — in mosses, said of a leaf **costa** that doesn't reach the **apex** (•below) (*see* **percurrent**).

Fissidens asplenioides **failing** costa. 100 µm

Bryobeckettia bartlettii **failing** costa. 100 µm

falcate — resembling a sickle blade, with edges parallel and lightly curved (•below and next page).

Pseudotaxiphyllum falcifolium **falcate** leaf. 50 µm

falcate-secund — having leaves that are strongly curved and turned toward only one side of the stem (•below, left and middle).

falciform — sickle-shaped (•below, right).

Ptilium crista-castrensis **falcate-secund** leaves. 1 mm

Hypnum lindbergii **falcate-secund** leaves. 1 mm

Warnstorfia fluitans **falciform** leaves. 1 mm

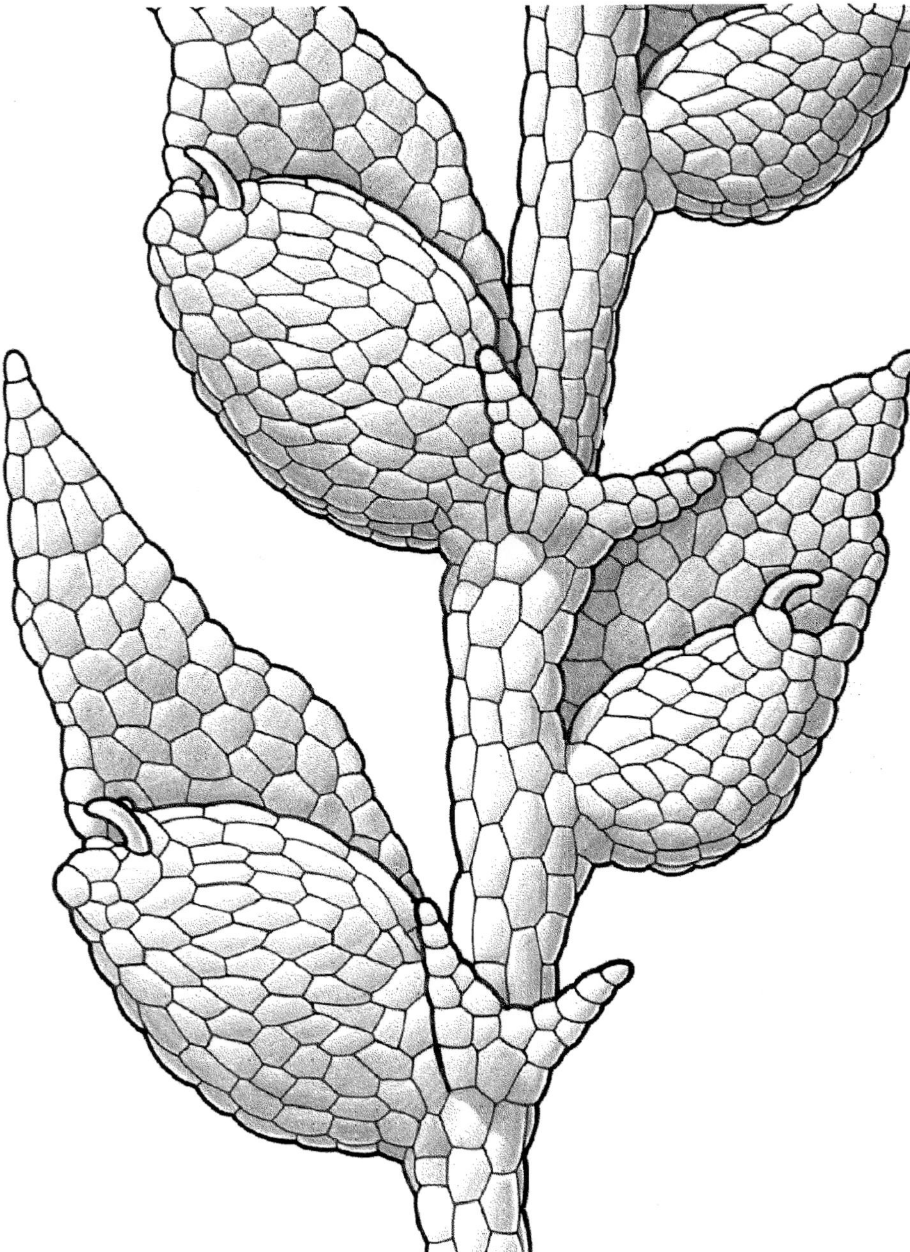

Drepanolejeunea aucklandica **falcate** lobule tooth. 0.1 mm

farinaceous — covered with small floury or mealy particles.
farinose — having the texture of flour.
fascicle — a bundle or cluster of branches, leaves, **propagules**, or other structures.
fasciculate — arranged in bundles (**fascicles**), bunched or bundled together.
fastigiate — having parts that are **erect**, nearly parallel, and almost the same length.
feather moss — a common name given to soft and matted, highly branched mosses like *Hypnum* (•below, left). Such mosses have long been used to stuff bed-pillows—*hypnos* means "slumber" in Greek.
felted — matted or tangled (•below, right).

Hypnum cupressiforme **feather moss.** 1 cm

Polytrichum juniperinum **felted** calyptrae. 1 mm

fenestrate — pierced by broad openings resembling windows (*compare with* **perforate**, having pores or small holes).
fern moss — the common name given to the moss genus *Thuidium* because its shoots resemble finely divided fern fronds (•below, left).
ferrugineous (or **ferruginous**) — rust-colored (•below, right).
fertile — producing sex organs (**antheridia** or **archegonia**) (*the opposite of* **sterile**).
festoon — a hanging loop or chain.

Thuidium furfurosum **fern moss.** 1 cm

Paracromastigum fiordlandiae **ferrugineous** leaf lobe. 50 μm

fibrillose — ornamented with fine spiral or **annular** fiber-like wall thickenings.

fibrils — fine spiral or **annular** fiber-like wall thickenings, as on the walls of **hyaline** cells (**leucocysts**) in the leaves and stem **cortex** of *Sphagnum* species (•below, right).

-fid — a suffix meaning *lobed* or *parted*—**bifid** means *two-lobed*, **trifid** *three-lobed*, and **quadrifid** *four-lobed* (•below, left).

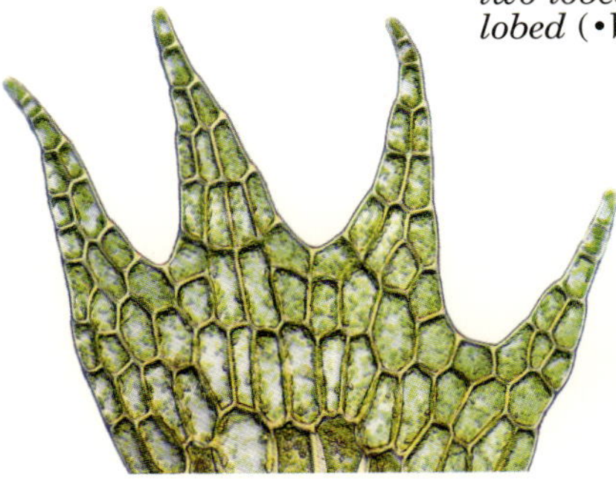

Telaranea patentissima quadri**fid** leaf apex. 100 µm

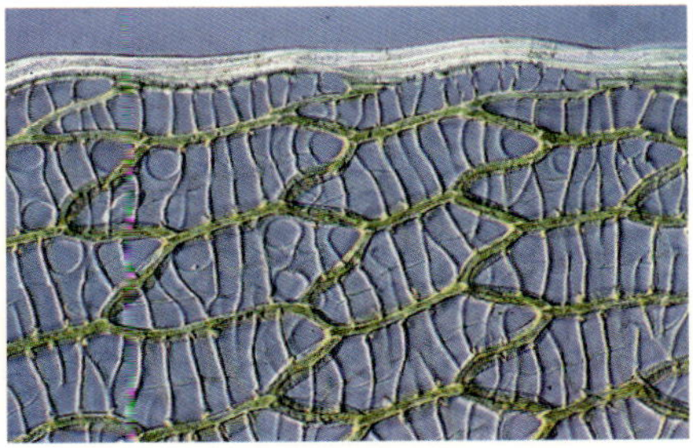

Sphagnum squarrosum leucocyst **fibrils**. 50 µm

filament — a single row of cells attached end-to-end, long and sometimes branched (•below).

Filaments (from left) *Achrophyllum dentatum* filamentous brood-body, *Leptotheca gaudichaudii* brood filament, and *Tetraphidopsis pusilla* filamentous gemma (various magnifications).

filamentous — resembling a thread (•below).

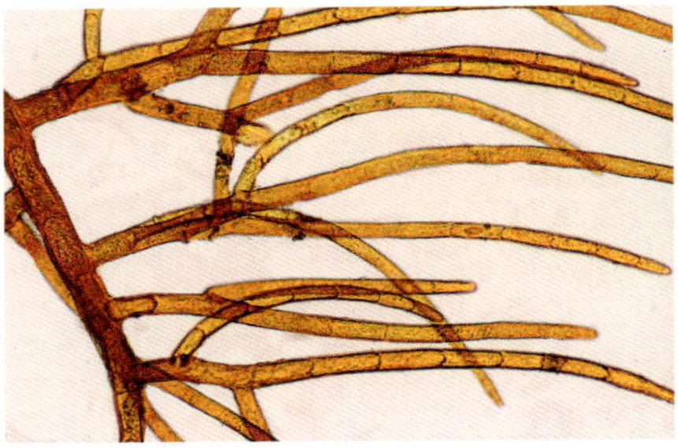

Ephemeropsis trentepohlioides **filamentous** protonematal gametophyte. 50 µm

Weissia controversa **filamentous** chloronemata and caulonemata. 50 µm

filiform — **filamentous**, thread-like, long and slender (•below) (*compare with* **capillary**, hair-like and therefore more delicate).

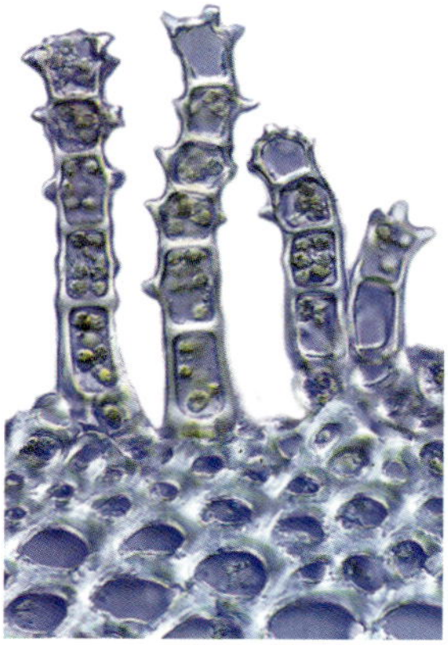

Thuidium laeviusculum **filiform** paraphyllia. 10 μm

Thuidium kanedae **filiform** paraphyllium. 50 μm

fimbriate — **fringed** (•below, left and middle) (*see* **hairy** for a comparison of similar terms).
fimicolous — growing on manure or dung.
fistulate — cylindrical and hollow but closed at both ends.
flabellate (or **flabelliform**) — fan-shaped (•below, right).

Campylopus pyriformis **fimbriate** calyptra base. 0.5 mm

Diphyscium foliosum **fimbriate** perichaetial leaf apex. 50 μm

Catharomnion ciliatum **flabellate** fronds. 5 mm

flaccid — soft, flabby, or limp.
flagellate — equipped with one or more **flagella**.
flagelliform — long, tapering, and supple like a whip or **flagellum**.
flagelliform branch — a slender, whip-like stem or branch, leafless or with much **reduced** leaves, which functions as an anchor or as a **propagule** (•left).
flagellum (plural **flagella**) — (1) a long thread-like organelle that protrudes from a cell, and functions in motility or feeding. The flagella of bryophytes and other eukaryotes are built of microtubules in a characteristic 9+2 array. They can vibrate but not rotate, and so they're termed whiplash flagella. Bryophyte sperms (**antherozoids**) have two such flagella, which can efficiently propel them through a thin film of water, (2) a **flagelliform branch**.
flavonoid — a group of plant pigments in which two 6-carbon rings are linked together by a 3-carbon unit. They include the water-soluble red and blue **anthocyanin** pigments and the fat-soluble yellow and orange carotenoid pigments. Flavonoids are widely distributed in liverworts and mosses but not hornworts. **Thallose** liverworts in the order Marchantiales produce diverse flavone glycosides. The nearly **cosmopolitan** moss *Hedwigia ciliata* produces at least a dozen different flavonoids. Although red pigments are common in liverworts and mosses, they're not usually anthocyanins, which have been reported with certainty from only two moss genera, *Splachnum* and *Bryum*.
fleshy — thick, pulpy, soft, and juicy (•below) (*synonym* **succulent**).

Lepidozia procera **flagelliform branch**.
100 μm

Aneura orbiculata **fleshy** thallus.
1 cm

Haplomitrium ovalifolium **fleshy** shoot.
1 mm

flexuose — zigzag or slightly wavy, bent, or twisted (•below).

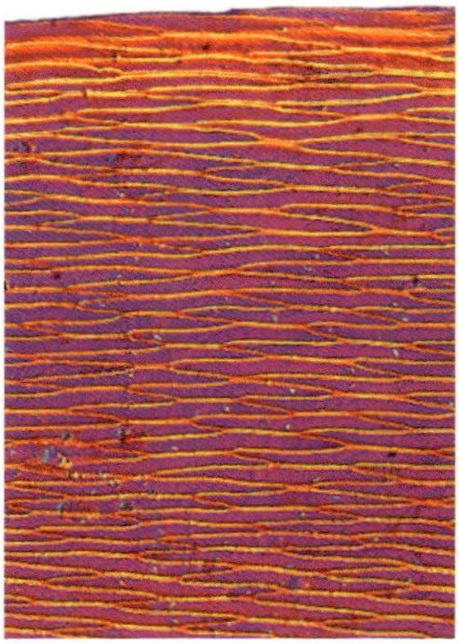

Dichelyma japonicum **flexuose** leaf cells. 100 μm

Bartramia pomiformis var. *elongata* **flexuose** leaves. 1 mm

Blindia lewinskyae **flexuose** leaves. 1 mm

Lopidium concinnum costa **flexuose** near apex. 100 μm

Pseudoscleropodium purum linear-**flexuose** leaf cells. 50 μm

Hypopterygium rotulatum **flexuose** underleaf costa. 100 μm

floccose — densely covered with hairs that fall in **tufts** (*see* **hairy** for a comparison of similar terms).

flora — the vegetation of a country, continent, or other selected area (*compare with* **Flora**, a formal **systematic** treatment of a flora).

Flora — a **systematic** treatment of a **flora** (*compare with* **flora**, the vegetation of a particular area).

floristic — referring to the **flora** of an area.

flux — the rate of flow of particles, energy, or liquid, for example the rate at which the energy of sunlight passes through the upper atmosphere.

foliate — leafy.
foliicole — a plant which grows on the living leaves of another plant (•below, left) (*see* **epiphyll**).
foliicolous — growing on leaves (*synonym* **epiphyllous**) (•below, left) (*see* **epiphyll**).
foliose — (1) leaf-like, (2) leafy (•below, middle).
foot — the base of the **sporophyte** that's embedded in the **gametophyte** and extracts water, mineral nutrients, and **photosynthate** (•below, right).

Echinolejeunea papillata **foliicolous** on *Blechnum chambersii* (fern).
1 mm

Rhizogonium distichum **foliose** habit.
1 mm

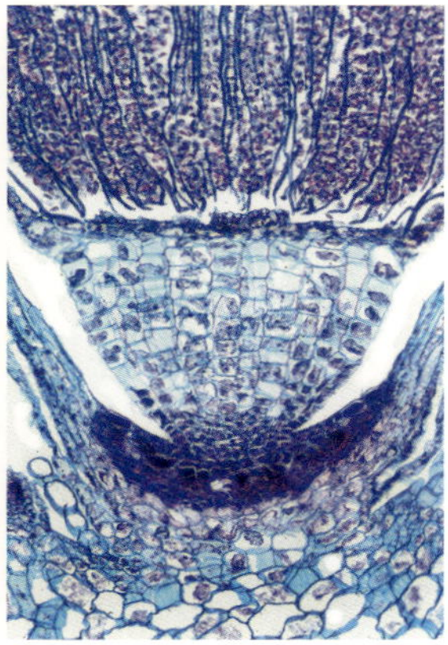

Mature liverwort sporophyte **foot** (ls).
100 μm

forcipate — curved inward like pincers or nippers.
forked moss — a common name given to *Dicranum* mosses because their **peristome teeth** are **forked**.
fornicate — (1) arched, (2) covered with **scales**.
foveate — pitted.
foveolate — delicately pitted like a golf ball.
fragmentation — the disintegration or breakup of a structure (•below). In many bryophytes, the fragments act as **diaspores**, **propagules**, or **disseminules**, sprouting into new plants.

Drucella integristipula **fragmenting** shoot. 100 μm

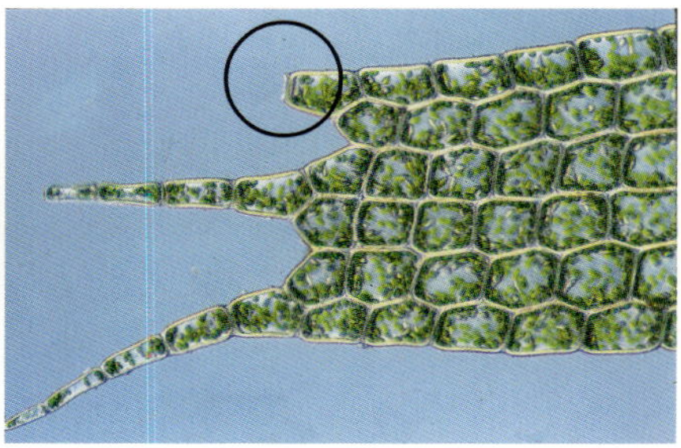

Telaranea fragilis **fragmenting** leaf lobe. 100 μm

fringed — having a short-**ciliate margin** or edge (•below, top left) (*synonym* **fimbriate**).

frond (adjective **frondose**) — a densely branched **shoot** system that's flattened into a single plane and resembles a fan, umbrella, or fern (•below, except for top left).

***Frullania*-type branching** — a type of branching in leafy liverworts in which a branch replaces the **postical** half of a leaf (for a comparison with other types of branching, *see* ***Acromastigum*-type branching**).

fugacious — (1) fleeting, (2) readily falling away.

fulvous — colored dull, tawny, yellow-brown.

fundus (plural **fundi**) — the base of a structure.

Distichophyllum pulchellum **fringed** calyptra.
0.5 mm

Hypnum chrysogaster **frondose** shoot.
1 mm

Hypnodendron menziesii **frondose** shoot.
5 mm

Herzogianthus vaginatus **frond**.
1 mm

Catharomnion ciliatum **frondose** shoot.
1 mm

Hypnodendron colensoi **frondose** shoot.
10 mm

funiform — rope-like (•below).

Myuroclada maximowiczii **funiform** shoots.
5 mm

Papillaria crocea **funiform** shoots.
10 mm

Lembophyllum divulsum **funiform** shoots.
10 mm

furcate — forked (•below) (*compare with* **bifurcate**, forked into roughly equal halves).

Rhizomnium striatulum **furcate** costa.
100 µm

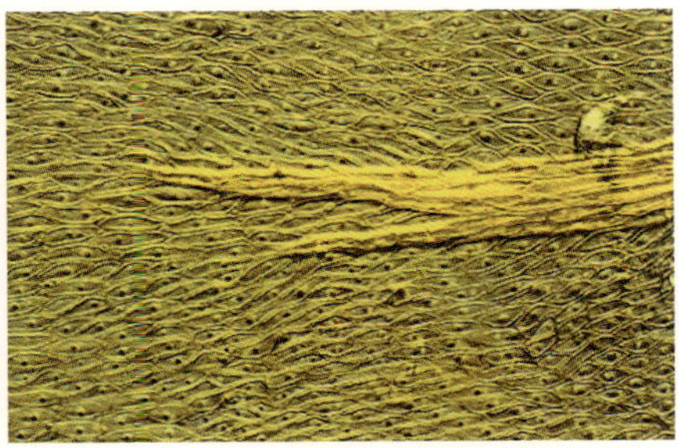

Aerobryopsis subdivergens **furcate** costa. 100 µm

Bryonoguchia molkenboeri **furcate** paraphyllia. 50 µm

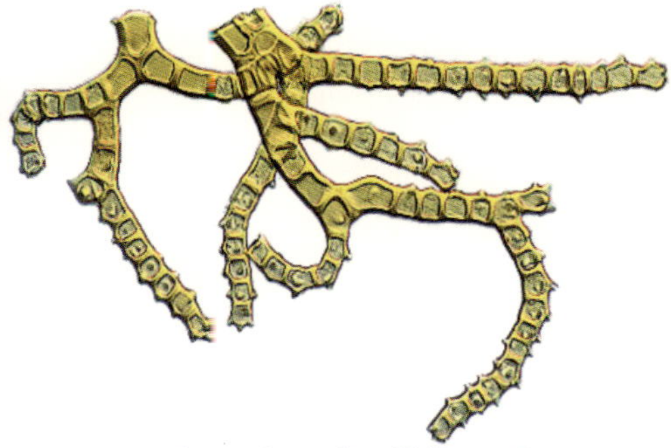

Thuidium kanedae filiform **furcate** paraphyllia. 50 µm

furfuraceous — scaly (•below, left).
furrowed — **grooved** (•below, middle and right).

Thuidium furfurosum
furfuraceous stem.
10 mm

Funaria hygrometrica
furrowed capsule (dry).
1 mm

Hypnodendron comatum
furrowed capsule (dry).
1 mm

fuscous — colored a dull dark-brown that's often tinged with black (•below, left).
fusiform — narrow and tapered at both ends (•below, middle and right) (*synonym* **spindle-shaped**).

Isotachis montana
fuscous shoots.
1 mm

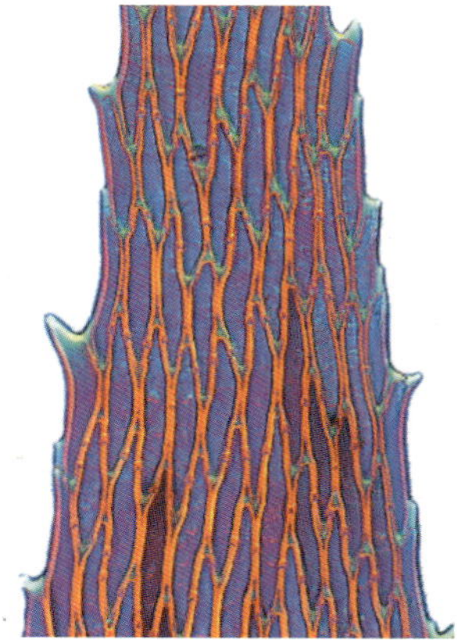

Garovaglia elegans
fusiform acumen cells.
100 μm

Sauloma tenella
fusiform leaf cells.
50 μm

galeate — helmet-shaped.
gametangium (plural **gametangia**) — a sex organ that forms **gametes**, either an **antheridium** that produces **sperms** (**antherozoids**) or an **archegonium** that produces **eggs** (**ova**).
gamete — a sex cell, either an **egg** (**ovum**) or a **sperm** (**spermatozoid**) and usually **haploid**.

Macromitrium longipes **gametophores**.
1 mm

gametogenesis — the production of sex cells (**gametes**).

gametophore — the leafy stem of the **gametophyte**, in mosses produced from **buds** on the **caulonema** stage of the **protonema** and usually bearing sex organs later (**gametangia**) (•left).

gametophyte — the **multicellular gamete**-producing generation of an **embryophyte**. It begins as a **haploid spore** shed by the **sporophyte**. In most mosses, it germinates and grows into a **filamentous protonema**, which then produces **buds** that grow into leafy plants with **rhizoids**. In liverworts, each protonema produces only a single plant. Hornworts lack protonemata. At maturity, the gametophyte produces sex organs, which eventually shed **gametes** (**sperms** or **antherozoids**) and **eggs** (**ova**). The gametophytes of most bryophytes are large and long-lived, whereas the sporophytes are small and short-lived (*see* **sporophyte**).

gasteropodous — shaped like a stomach, bulging or swollen on one side toward the base (*similar to* **ventricose** and **strumose**, swollen goiter-like on one side of the base) (*compare with* **gibbous**, bulging or swollen on one side toward the top).

gelatinose — resembling jelly.

geminate — arranged in pairs (•below, left).

gemma (plural **gemmae**) — a type of **vegetative propagule**. Composed of only a few cells, gemmae can look like **buds**, leaves, or branches, and they can be formed on thalli, stems, **rhizoids**, or leaves (or in leaf **axils**) or on specialized structures. Under favorable conditions, one cell **differentiates** into an **apical** cell which in turn develops into a new **thallus** or leafy **gametophore** (•below, middle and right).

Trachycystis flagellaris **geminate** marginal teeth.
0.5 mm

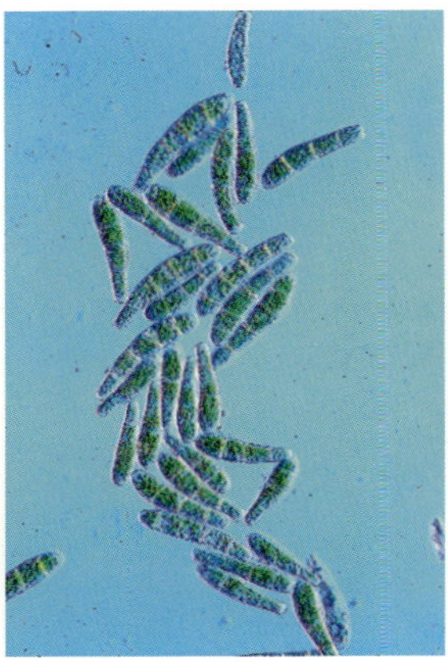

Tetraphidopsis pusilla **gemmae**.
100 μm

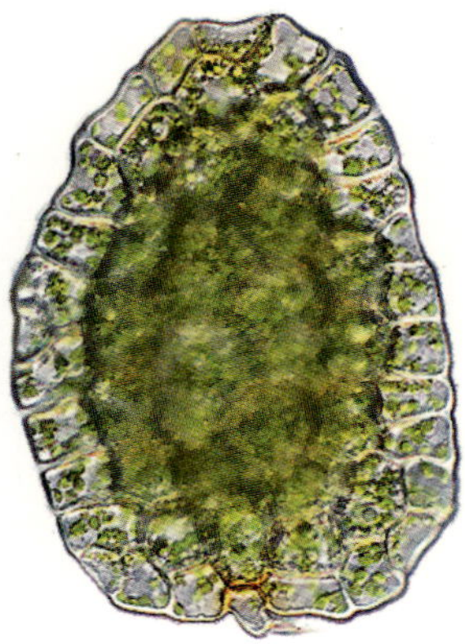

Tetraphis pellucida disciform **gemma**.
50 μm

gemmae cup — a **gemma**-producing structure shaped like a cup or crescent on the upper surface of some **thallose** liverworts (•below, left).

gemmiferous (or **gemmiparous**) — bearing **gemmae** or buds.

gemmiform (or **gemmate**) — **bud**-like.

gene — a portion of a DNA molecule that controls the synthesis of a polypeptide, a molecule made up of amino acids linked end-to-end in a long chain. The information on the sequence of amino acids in the chain is coded along the DNA chain in the form of triplets of base-pairs (either adenine with thymine or cytosine with guanine). The entire gene is called a cistron. Its coding regions (called exons) are preceded by a leader and followed by a trailer, and usually are broken up by several non-coding regions called introns. When the gene is expressed, it's first transcribed into RNA. The non-coding introns are transcribed along with the coding exons, but they're soon snipped out and the exons stitched together. The RNA travels to the cytoplasm where it's translated into a polypeptide by ribosomes.

Warnstorfia fluitans **geniculate** setae.
1 mm

generalized — said of an organism that's thought to resemble most closely the earliest ancestor (progenitor) of its evolutionary lineage (*the opposite of* **derived** or **specialized**).

geniculate — **abruptly** bent like a knee (•left).

genus (plural **genera**) — (1) a group of related **species**, and thus (2) the unit of **taxonomic** classification just above the **species**, (3) the first of the two words in a Latin binomial, a scientific name for a species, such as *Andreaea* in the binomial *Andreaea rupestris*.

Marchantia foliacea **gemmae cups**.
5 mm

Bazzania adnexa positively **geotropic** flagelliform root. 1 mm

geotropic — growing directionally in response to gravity—downwards growth is termed positive, and upwards is termed negative (•above, right).

geotropism (or **gravitropism**) — a growth response of a plant that's induced by the force of gravity.

germ cells — the reproductive cells of an organism (*compare with* **somatic cells**, the non-reproductive cells of the body of an organism).

germination — the beginning of growth of a **spore** or other previously dormant structure.

germling — the mass of unspecialized cells giving rise to the **apical** cell in a liverwort **gametophyte**.
germ tube — the **filament** that emerges first from a germinating **spore**.
gibbous — bulging or swollen on one side toward the top (*compare with* **ventricose**, **strumose**, and **gasteropodous**, bulging **basally**, stomach-like).
gigas — very large (•below, left).
glabrous — smooth (•below, middle and right).

Dicnemon dixonianum **gigas** multicellular spore. 100 µm

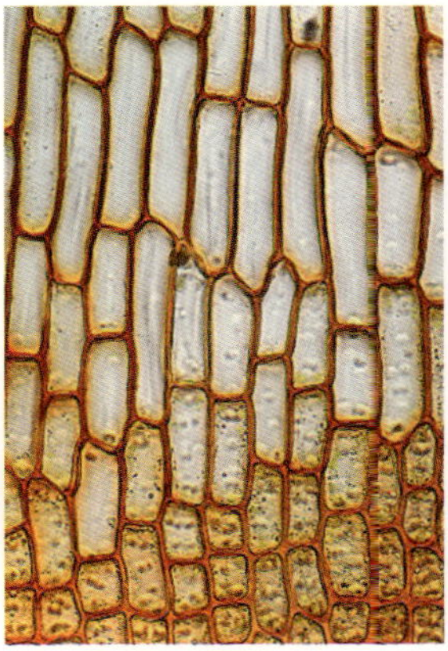

Tortula muralis **glabrous** leaf base cells. 100 µm

Marsupidium knightii **glabrous** leaf surfaces. 1 mm

glacier mice — (*see* **moss balls**).
glaucous — having a whitish or grayish **bloom** like the surface of a plum (•below, left and middle) (*compare with* **caesious**, which is more bluish).
globose — spherical or nearly so (•below, right).

Anthelia juratzkana **glaucous** shoots. 1 mm

Philonotis scabrifolia **glaucous** stem leaves. 0.5 mm

Conostomum pentastichum **globose** capsules. 1 mm

Gondwana (or **Gondwanaland**) — an ancient supercontinent which broke up into present-day India and the Southern Hemisphere continents.

gonidium (plural **gonidia**) — **gemma**.

gonioautoicous — **monoicous**, with the male **bud**-like in an **axil** of the same stem or branch as the female **inflorescence** (for a comparison with other arrangements, *see* **autoicous**).

gonium (plural **gonia**) — a group of **antheridia** plus their **sterile paraphyses**.

granular — said of moss leaf **costae** which are roughened **adaxially** by projecting cells.

granulose — (1) minutely grainy, (2) roughened with minute blunt projections.

gravitropism — a plant growth response induced by the force of gravity (*synonym* **geotropism**).

greenhouse moss (or **glasshouse moss**) — a common name of the greenhouse "weed moss" *Leptobryum pyriforme*, which can readily spread over the bare soil of flowerpots because it produces copious **brood bodies** on its **rhizoids**. The **cosmopolitan** liverworts *Marchantia polymorpha* and *Lunularia cruciata* (•below, middle) are also notorious greenhouse weeds, spreading by **gemmae**.

gregarious — growing close together but not very densely, as in **mats** or **tufts** (•below, right).

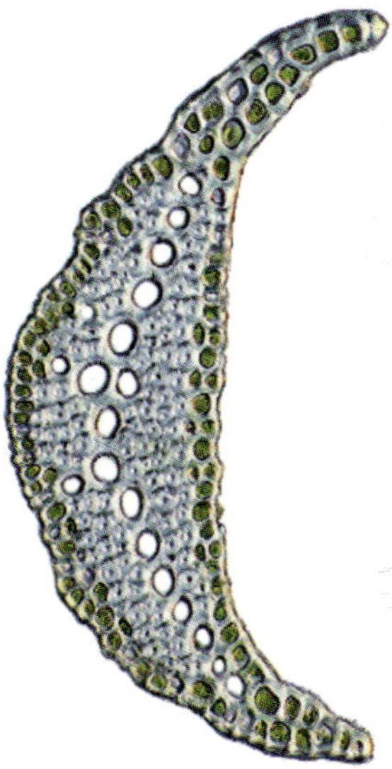

Echinodium hispidum **guide cells** (leaf xs).
50 µm

Lunularia cruciata **greenhouse** weed.
5 mm

Tortula truncata **gregarious** habit.
5 mm

grooved — **channeled** (*synonym* **furrowed**).

guard cells — the specialized inflatable epithecial cells surrounding a moss or hornwort **stoma**, usually paired and kidney-shaped, but in some species of the moss family Funariaceae only a single cell with a central slit-like opening.

guide cells — in some mosses, large, empty, thin-walled cells stretching across the **costa** (•above, left) (*synonym* **deuters**).

guttulate — (1) having cell **lumina** with one or more oil-like droplets, (2) having lumina appear rounded from heavily thickened walls (•below, right).

gymnostomous — in mosses, lacking a **peristome** (*gymnostome* means *naked mouth* in Greek).

gynoecial branch — a specialized branch bearing **gynoecia** (•below, left).

Dendromastigophora flagellifera **gynoecial branch**. 1 mm

Schistidium apocarpum **guttulate** basal leaf cells. 25 μm

gynoecium (plural **gynoecia**) — the "female house", hence the **archegonia** and **paraphyses** (if present) plus any surrounding modified leaves.

gyrate — curved into a circle (•below, left) (*similar to* **circinate**).

gyrothecal — said of branches below the tip (*the opposite of* **athecal**) (*similar to* **intercalary**).

habit — general appearance (*synonym* **facies**).

habitat — the type of environment an organism typically lives in (•below, middle).

hair cap moss — a common name given to the family Polytrichaceae because their **calyptrae** are densely **hairy** (•below, right).

Ptilium crista-castrensis **gyrate** leaves. 1 mm

Forest **habitat**.

Polytrichum sp. **hair cap moss**. 1 mm

hair-point — in mosses, a hair-like and often **hyaline** leaf tip, formed either by a **costa**'s projecting well beyond the end of the leaf blade, or by a protracted tapering of the blade tip. Hair-points often turn white when they dry out, making the moss look frosted (*synonym* **hoary**) from a distance. Mosses adapted to **habitats** that periodically dry out often have hair-points, and some can minimize water loss by controlling the orientation of their hair-points. *Campylopus introflexus* bends its hair-points down roughly 90° during dry periods, forming a dense mat that traps a thick "boundary layer" of quieter air, thus markedly lowering water loss (•below and top of next page, left and middle).

Campylopus introflexus **hair-points**.
10 mm

Leptostomum inclinans **hair-points**.
5 mm

Racomitrium curiosissimum **hair-points**.
100 µm

Campylopus introflexus control of hair-point orientation: erect when hydrated (left), reflexed when desiccated (middle and right).

Leptostomum macrocarpon **hair-points**.
1 mm

Leptostomum macrocarpon **hair-point** (base, wm).
100 μm

Dicranum sp. **haplolepidous** peristome.
1 mm

hairy — covered with hairs or producing hairs.

COMPARISON OF HAIRINESS TERMS:

arachnoid — covered with white, entangled hairs.
barbate — bearded with tufts of long hairs.
ciliate — having fine hairs on the margin.
comose — hairy.
echinate — roughened by bristly hairs or prickles.
fimbriate — fringed.
floccose — covered with hairs that fall off in tufts.
hispid — covered with stiff hairs.
lanate — woolly.
pilose — clothed with hairs.
plumose — feathery.
pubescent — downy.
sericeus — densely covered with silky hairs.
strigose — covered with rigid appressed hairs.
velutinous — velvety.
villose — shaggy, covered with long hairs.

hamate (or **hamulose**) — shaped like a hook (*synonyms* **aduncous**, **hooked**, and **uncinate**).

haploid — said of an organism which contains only one complete set of **chromosomes** (denoted by **n** or 1n) in the nuclei of its cells—typically, bryophyte **gametophytes** are haploid (*compare with* **diploid**, said of a cell which contains two complete sets of chromosomes (denoted by 2n)—typically, bryophyte **sporophytes** are diploid) (*see* **ploidy level**).

haplolepidous — in mosses, (1) said of a **peristome** with only one ring of **teeth** (• above, right) (*the opposite of* **diplolepidous**) (*synonym* **single peristome**), (2) said of a jointed (**articulated**) peristome tooth in which the outer face of each joint is formed from a single cell.

hapteron (plural **haptera**) — a swollen tip where a **protonema** or **rhizoid** attaches to a **substratum**.

hastate — arrow-shaped, **abruptly** broadened at the base into **auricles**.

haustorium (plural **haustoria**) — a structure at the base of the **sporophyte** that penetrates the **gametophore** and absorbs water, mineral nutrients, and energy-rich **photosynthate** (•below, left). That's important for the development of the **capsule** and the **spores** inside it, because as the sporophyte ages, it steadily loses chlorophyll and therefore increasingly depends on its parent **gametophyte** for its supply of organic nutrients.

hebetate — having a soft, blunt tip.

helical (or **helicoid**) — twisted or shaped like a spiral, helix, or the thread of a screw (•below, middle and right) (*see also* **dextrorse** and **sinistrorse**).

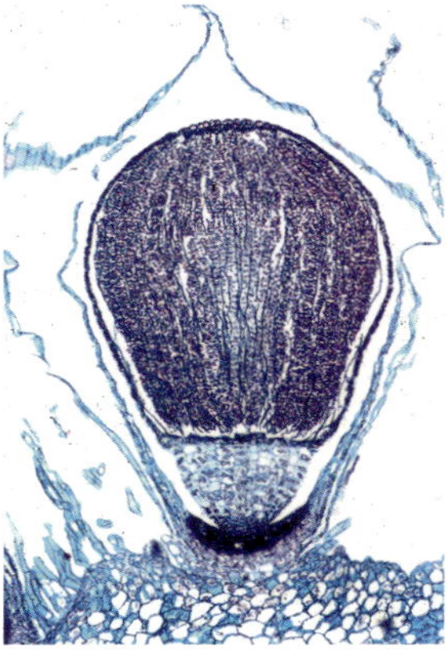

Liverwort sporophyte **haustorium** (section). 100 µm

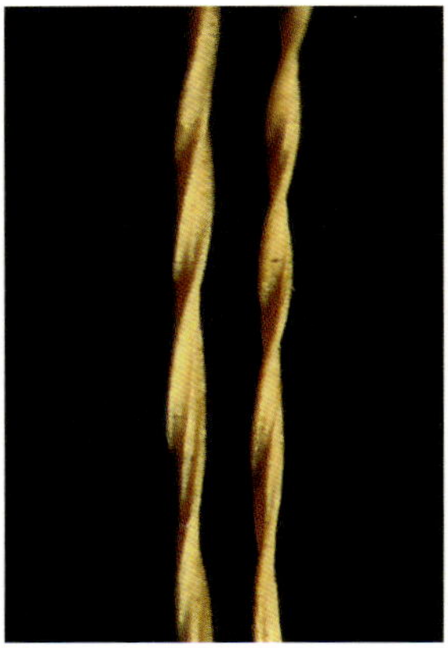

Entodon rubicundus **helical** setae (dry). 1 mm

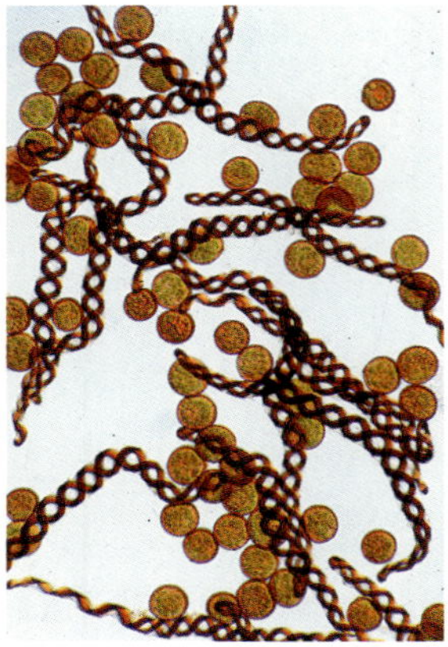

Adelanthus falcatus **helical** elaters. 100 µm

heliophilous — thriving in a sunny **habitat** (*synonym* **photophilous**).

heliophyte — a plant adapted to full sunlight.

hemi-isophyllous — having stem leaves and branch leaves which are similar in their size, shape, and structure.

heterochromatin — regions of **chromosomes** that remain heavily condensed, and so their genes are not expressed (translated into protein).

heterogeneous — made up of contrasting or unrelated parts (*the opposite of* **homogeneous**, uniform or composed of similar parts).

heteroicous — said of a **species** that produces various types of **inflorescences** on different plants or on the same plant, so that it confusingly seems to be both **dioicous** and **monoicous** (*synonyms* **polygamous** and **polyoicous**).

heteromallous — pointing in various directions (*compare with* **homomallous**, pointing in the same direction).

heteromorphous — having different shapes.

heterophyllous — producing clearly different types of leaves on stems and branches (*the opposite of* **isophyllous**, producing the same type of leaf on stems and branches).

heterothallic — said of a **species** that can't self-fertilize because its **gametes** aren't compatible (*the opposite of* **homothallic**, self-fertile).

heterotrichous — having different types of hairs.

hexagonal — having six nearly equal sides (•below, left). Unequally six-sided cells are described with modifiers such as **unequally** hexagonal (•below, middle) or **oblong**-hexagonal (•below, right).

Calyptrochaeta cristata **hexagonal** leaf cells. 100 μm

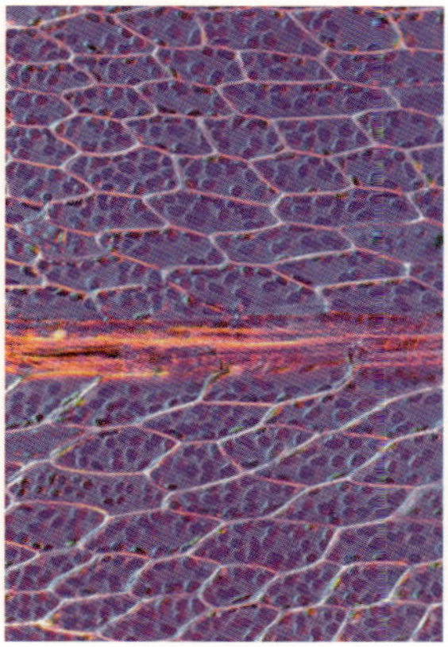

Bryum argenteum unequally **hexagonal** leaf cells. 50 μm

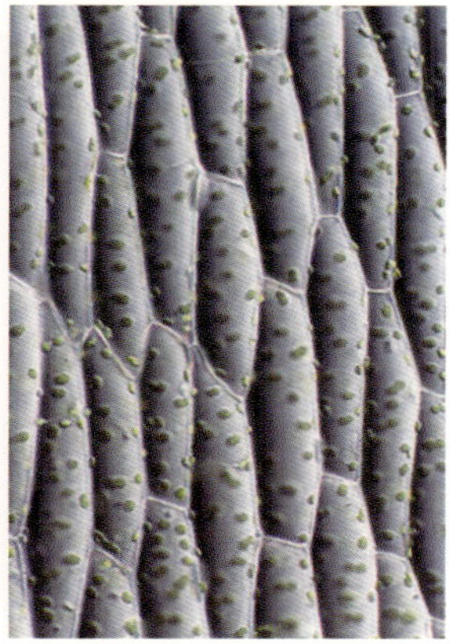

Bryum billardierei oblong-**hexagonal** leaf cells. 50 μm

hispid — rough or bristly, covered with stiff hairs (*see* **hairy** for a comparison of similar terms).

hoary — appearing frosted, whitish, or grayish from massed **hyaline** leaf tips (•below).

Campylopus introflexus **hoary** massed hair-points. 1 cm

Racomitrium curiosissimum **hoary** shoot tips. 1 mm

holarctic — a distribution pattern in which a **species** is widely distributed throughout the Northern Hemisphere.

holostipous — said of **entire underleaves**.
homogeneous — uniform or composed of similar parts (•below) (*the opposite of* **heterogeneous**, made up of contrasting or unrelated parts).

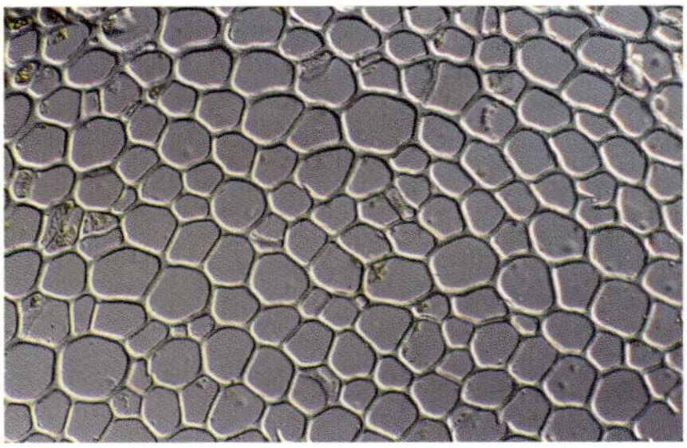

Plagiochila sp. **homogeneous** stem cells (xs). 100 μm

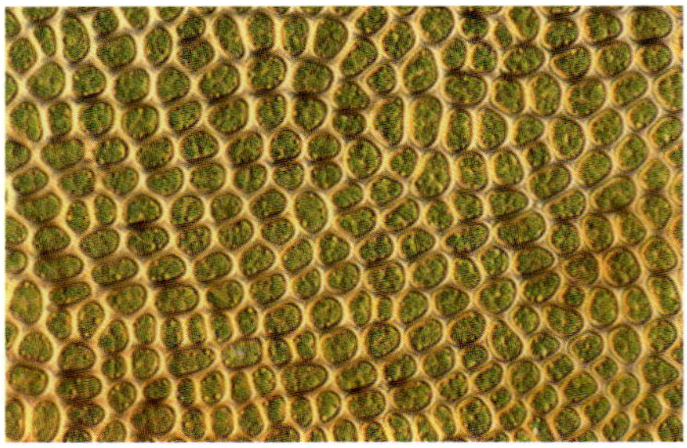

Tetraphis pellucida **homogeneous** leaf cells (wm). 100 μm

homologous (noun **homology**) — said of two organs or structures that have the same phylogenetic origin but not necessarily the same function or appearance (*the opposite of* **analogous**, said of two organs or structures having the same function but a different phylogenetic origin).
homomallous — all pointing in the same direction (*compare with* **heteromallous**, pointing in various directions).
homothallic — said of a **species** that's self-fertile (*the opposite of* **heterothallic**, said of a species that can't self-fertilize because its **gametes** aren't compatible).
hooked — hook-shaped (•below) (*synonyms* **aduncous**, **hamate**, and **uncinate**).

Bryum argenteum **hooked** seta. 1 mm

Hypnum lindbergii **hooked** shoot tip (dry). 1 mm

Bryum billardierei **hooked** seta. 1 mm

horizontal — parallel with the earth's surface or nearly so (•below).

Hypopterygium commutatum **horizontal** capsules. 1 mm

Rhizogonium distichum **horizontal** capsules. 1 mm

horn — the long tapering **sporophyte** of a hornwort (•below, left). Unlike the sporophytes of liverworts and mosses, the horn lacks a **seta** and its growth is indeterminate. It continues to grow as long as conditions are favorable, shedding mature **spores** from its tip and producing new spores from the **intercalary meristem** at its base just above the **foot**. The horns of most species **dehisce** along one or two longitudinal lines of weakness.

humic — rich in partly decayed vegetation and other organic matter (humus).

humicolous — growing on decomposing organic matter (humus).

humidity — dampness, or an inexact measure of how moist the air is.

hyalescent — somewhat colorless or transparent.

hyaline — colorless and transparent or nearly so, lacking chlorophyll or other pigments (•below, middle and right, and next page).

Anthoceros laminiferus mature **horns**. 1 mm

Polytrichum juniperinum **hyaline** limb cells. 50 µm

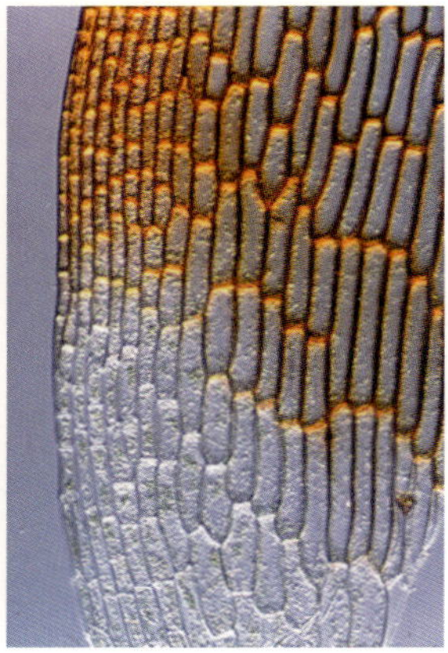

Tortula muralis **hyaline** basal marginal leaf cells. 100 µm

hyaline (continued)

Funaria hygrometrica **hyaline** tips of peristome teeth. 50 µm

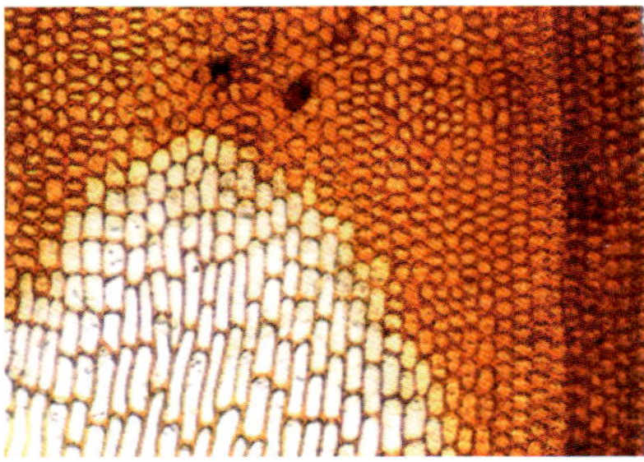

Tortula ruralis **hyaline** leaf base. 100 µm

hyaline margin — in liverworts, a leaf or **thallus** edge which lacks chlorophyll or other pigments and sometimes **oil-bodies** as well (•below, left).

hyalocyst — in the leaves of *Sphagnum* species, a colorless water-storage cell (•below, right) (*synonyms* **hyaline cell** and **leucocyst**).

Herzogobryum teres **hyaline margin**. 100 µm

Sphagnum squarrosum **hyalocysts** in leaf. 100 µm

hyaloderm (or **hyalodermis**) — in the stems of leafy liverworts and species of *Sphagnum*, a **cortex** composed of large, empty, colorless cells (•below).

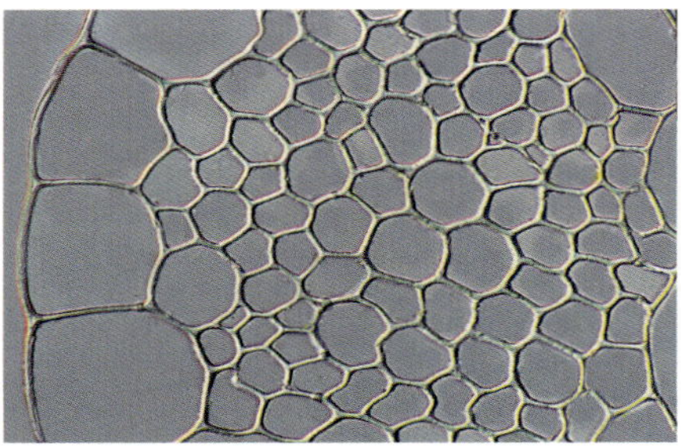

Telaranea gibbsiana **hyaloderm** (stem xs). 100 µm

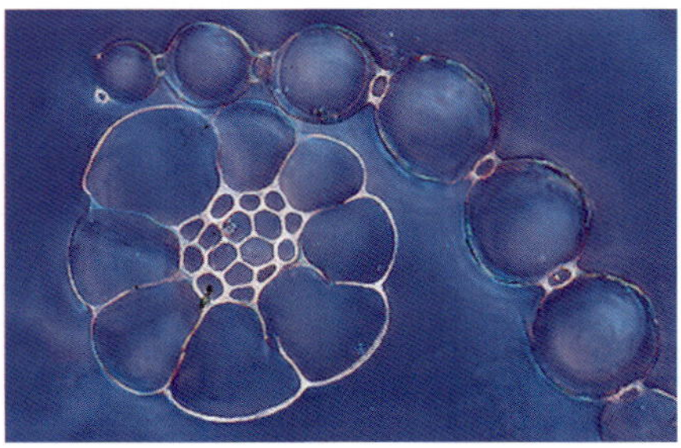

Sphagnum sp. **hyaloderm** (branch xs). 100 µm

hybrid — the offspring of two organisms belonging to different **species**.

hydrated — (1) wetted, (2) said of a plant that was dry but has been re-wetted enough to resume normal metabolism.

hydroids — specialized cells in the **central strand** and **costa** of some bryophytes. They conduct water and resemble the tracheids of vascular plants (•below) (*compare with* **leptoids**).

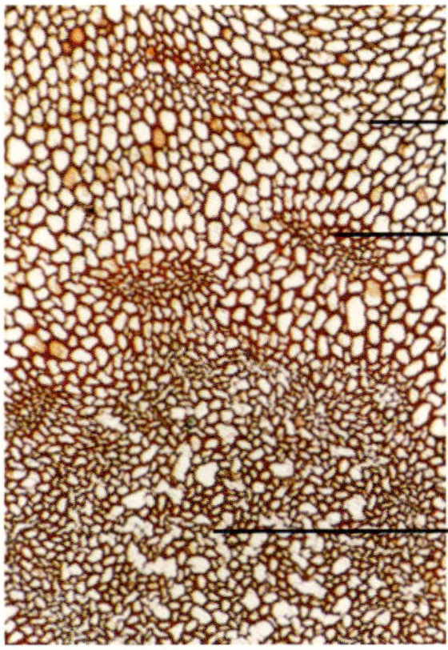

Dawsonia superba stem **hydroids** (xs).

100 μm

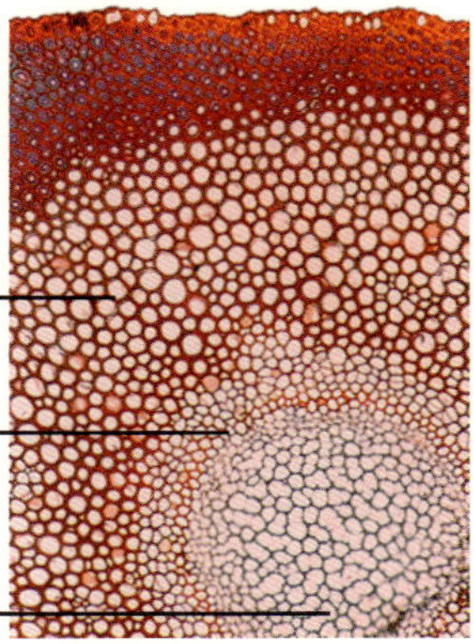

Dendroligotrichum dendroides stem **hydroids** (xs).

hydrome — a collective term for all of the water-conducting cells (**hydroids**) in a **central strand**.

hydrophilic — readily wetted.

hydrophilous — thriving in an **aquatic** environment, either submerged or floating (*compare with* **hygrophilous**, **mesophilous**, and **xerophilous**).

hydrophobic — resistant to wetting (•below).

Barbula unguiculata **hydrophobic** habit. 1 mm

Atrichum androgynum **hydrophobic** shoots. 1 cm

Adelanthus falcatus **hydrophobic** sporophyte. 1 mm

hydrophyte — a plant adapted to an **aquatic habitat**, growing submerged or floating (•below, left) (*compare with* **hygrophyte**, adapted to wet conditions but not growing in water (•below, right), **mesophyte**, and **xerophyte**).

hygrophilous — preferring a wet but not **aquatic habitat** (•below, right) (*compare with* **hydrophilous**, adapted to a true aquatic habitat, growing submerged or floating (•below, left), **mesophilous**, and **xerophilous**).

Blindia lewinskyae **hydrophyte**.
1 mm

Sphagnum sp. **hygrophyte**.
5 mm

hygrophyte — a plant adapted to wet **habitats** but not growing in water (•above, right) (*compare with* **hydrophyte**, a true **aquatic** growing submerged or floating in water, **mesophyte**, and **xerophyte**).

hygrophytic — (*see* **hygrophyte**).

hygroscopic (or **hygrometric**) — (1) readily absorbing moisture, (2) said of **moss peristome teeth** that respond to changes in humidity by bending in or out. The inner and outer surfaces of a peristome tooth differ in their responses to **humidity**. As a result, the tooth bends much as a bimetallic strip bends with changes in temperature. Many peristome teeth move instantly and dramatically with even fleeting and subtle changes in humidity, as if alive. **Single peristomes** usually have only a limited response—they bend outwards in dry air and inwards in moist air. In contrast, the movement of **double peristomes** is more complex—most bend inward when they're *either* bone-dry or soaked, so the **capsule** opens only during marked *changes* in humidity like the onset of a shower or weather clearing after a rain. A **spore** dispersed under those conditions is more likely to alight on a moist surface, ensuring that it will stick there and also have enough moisture to germinate.

Zoopsidella caledonica **hyoid** leaf sinus.
100 µm

hylacolous — growing on bark (*synonym* **corticolous**).

hylophilous — thriving in forest **habitats**.

hyoid — U-shaped or horseshoe-shaped (•left) (*synonym* **ypsiloid** or **ypsiliform**) (*compare with* **digammoid** or **digammiform**, V-shaped).

hypnoid peristome — a complete **double peristome**, with 16 **lanceolate teeth**, cross-**striolate** at the base and **trabeculate** at the back, and a well-developed **endostome** consisting of a high **basal membrane** and 16 lanceolate, **keeled segments** alternating with one or more **cilia** (•below, left and middle).

Cryptopodium bartramioides **hypnoid peristome**. 1 mm

Ptychomnion aciculare **hypnoid peristome**. 1 mm

Bryum argenteum shoot with **imbricate** leaves. 1 mm

hypogaeus (or **hypogaeous**) — growing below the surface of the soil.

hypophysis (plural **hypophyses**) — in mosses, the swollen **neck** of a **capsule** (**sporangium**) where it joins the **seta** (*synonym* **apophysis**).

IAA — an acronym which stands for indole acetic acid, a plant hormone that in mosses triggers the switch from the **chloronema** to the **caulonema** stage of **protonemata** (only the caulonema produces leafy **gametophores**) (*see* **auxin**).

imbricate — closely **appressed** and overlapping like shingles (•above, right, and below, left).

immarginate — lacking a **border** (•below, right).

Paraschistochila conchophylla **imbricate** leaves. 5 mm

Forsstroemia trichomitria **immarginate** leaf. 100 µm

immersed — hidden or submerged, (1) in mosses, said of **capsules** that are completely overtopped by the **perichaetial** leaves (•below) (*the opposite of* **exserted**, lifted beyond the tips of the perichaetial leaves) (*compare with* **emergent**, protruding only partly beyond the tips of the perichaetial leaves), (2) said of stomatal **guard cells** that lie below the level of the **exothecial** cells.

imperfect — said of a structure that lacks some of its usual parts.

Neckera pennata **immersed** capsule.
0.5 mm

Cryptopodium bartramioides **immersed** capsules.
1 mm

Schistidium apocarpum **immersed** capsule.
1 mm

inaperturate spore — a **spore** that has no "apertures" or thin spots in its outer wall.

incised — cut or dissected into sharp divisions.

inclined — angled 20°–60° from vertical (•below).

Psilopilum crispulum **inclined** capsules.
3 mm

Polytrichadelphus magellanicus **inclined** capsules.
2 mm

Leptostomum inclinans **inclined** capsules.
5 mm

included — enclosed or nearly so by surrounding organs or structures (•below, left).

inclusions — objects such as crystals enclosed within cells (•below, right).

Dendromastigophora flagellifera **included** sporophytes. 0.5 mm

Hypnodendron colensoi crystal cell **inclusions**. 100 μm

incoherent — loosely **tufted** and readily separating when collected.

incrassate — thick-walled (•below) (*compare with* **crassate**, thick).

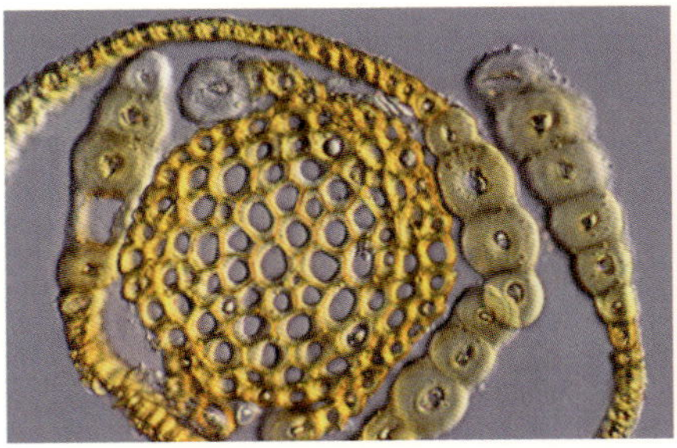

Dicnemon dixonianum **incrassate** leaf cells. 50 μm

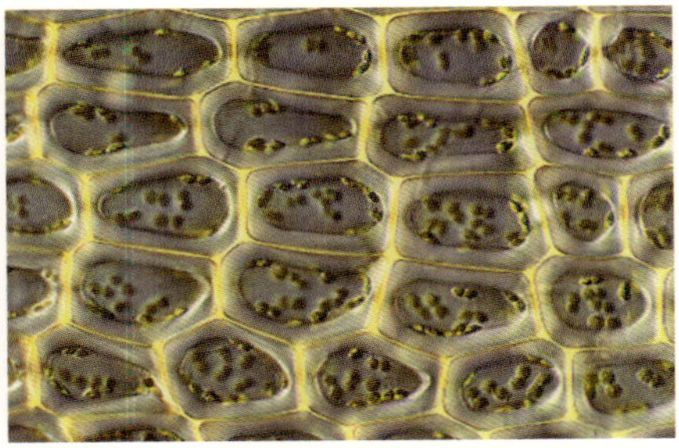

Leptostomum macrocarpon **incrassate** leaf cells. 50 μm

incubous leaf insertion — one of three ways that leaves are attached to stems in bryophytes. To visualize all three, imagine that you're holding the stem up at eye-height, that its tip is pointed toward the ceiling, and that you're looking at its upper surface (the side opposite any **underleaves**). If you can easily see the leading edges of the leaves (the edges facing the tip of the stem) but not their trailing edges (the edges facing the base), then the leaves are said to be **incubously inserted**. Conversely, if you can see the trailing edges more easily, then they're said to be **succubously inserted**. Lastly, if the leaves are nearly edge-on, they're said to be **transversely inserted** (•next page). Bryophyte leaves don't always fit those three categories neatly, but many do, and so leaf insertion is a useful tool for identification and classification.

incubous leaf insertion (continued) 108

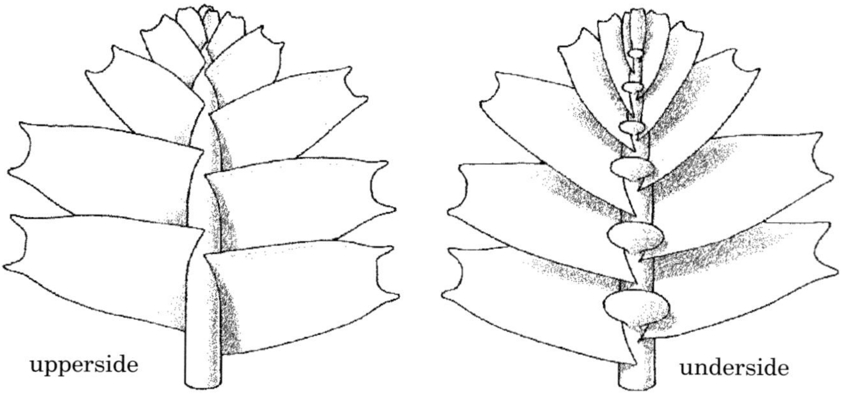

incubous leaf insertion

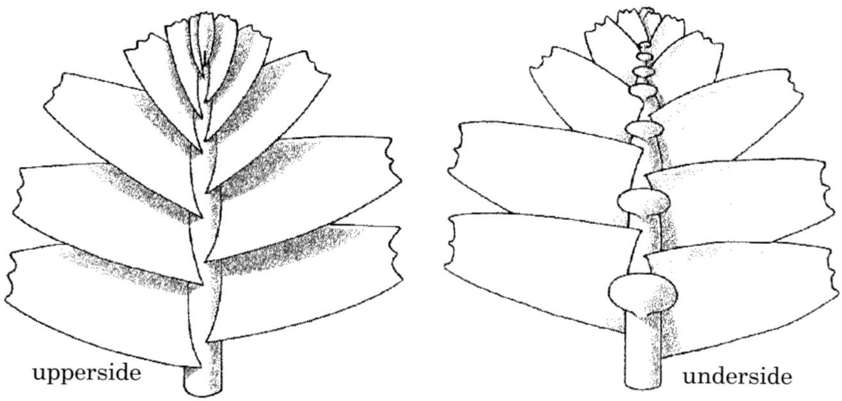

succubous leaf insertion

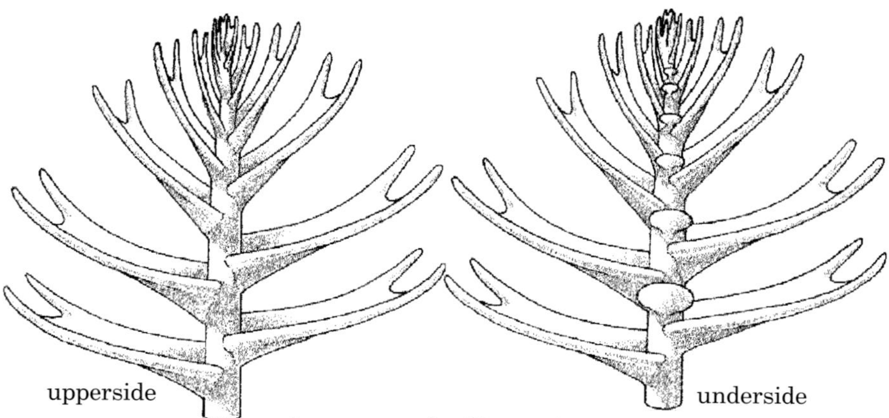

transverse leaf insertion

incumbent — lying or leaning on something.
incurved — curved upwards and inwards (•below).
indehiscent — said of a moss **capsule** that does not open along built-in lines of weakness such as **valves** or an **operculum**.

Barbula sp. **incurved** leaf margins (xs). 10 µm

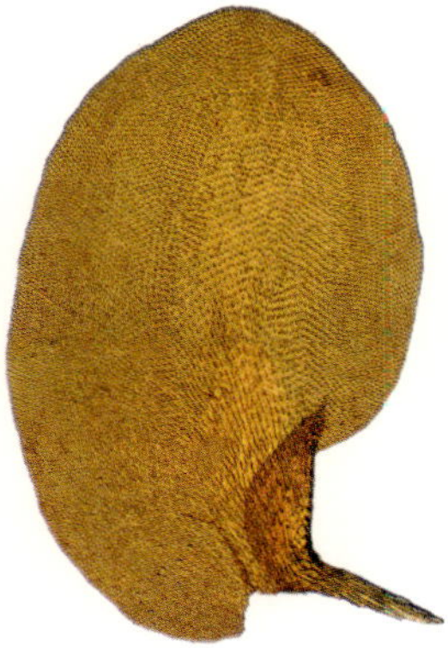

Homaliadelphus targonianus **incurved** leaf base. 0.5 mm

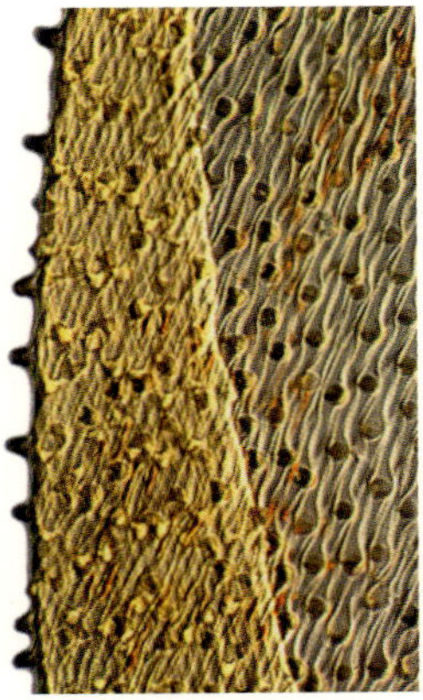

Rhaphidostichum boschii **incurved** leaf. 100 µm

indigenous — native (*the opposite of* **adventive**, introduced).
inflated — (1) swollen (•below) (*the opposite of* **explanate**, flattened), (2) membranous, transparent, and inflated like a bladder.
inflexed — **abruptly** bent upwards or inwards (*compare with* **reflexed**, abruptly bent or curved downwards and inwards).

Funaria hygrometrica **inflated** calyptra. 1 mm

Callicladium haldanianum **inflated** alar region. 100 µm

Pulchrinodus inflatus **inflated** leaves. 1 cm

inflorescence — in bryophytes, a cluster of sex organs plus any specialized leaves or **bracts** that surround them (•below, left) (*not the same as* **inflorescence** in the flowering plants, the pattern of how flowers are borne and/or ripen on a stem, such as *panicle* and *raceme*).

infundibuliform — funnel-shaped.

initial — an unspecialized cell that divides repeatedly, its descendants developing into specialized tissues and organs.

inner peristome — in mosses, the **endostome**, usually arising from a **basal membrane** and consisting of **segments** and **cilia** (*see* **peristome**).

innovation — (1) any new branch (•below, right), (2) in some mosses and liverworts, a **lateral** branch that forms at the base of or within an **inflorescence** after the sex organs have matured.

Hypopterygium rotulatum antheridial **inflorescences**. 1 mm

Calliergonella cuspidata **innovations**. 1 mm

inoperculate — in mosses, (1) said of a **capsule** after its **operculum** has fallen away (*synonym* **deoperculate**), (2) said of a capsule that has no operculum, and opens by the decay or breakdown of its wall (*synonym* **cleistocarpous**).

inrolled — rolled inwards (•below) (*synonym* **involuate**) (*compare with* **involute**, which denotes more tightly rolled).

Hylocomium splendens **inrolled** leaf margin. 100 μm

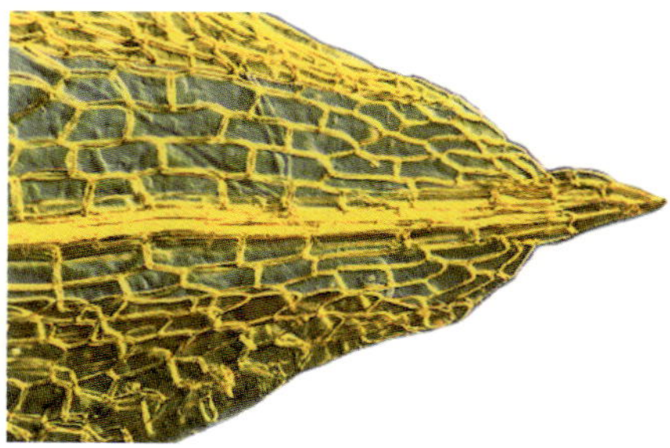

Funaria hygrometrica **inrolled** leaf margins. 50 μm

insertion — the line or point of attachment of a leaf or branch to a stem (*see* **incubous** for a comparison of insertion types in bryophytes) (•below).

interactions — the effects that living things have on each other, best measured as a boost or decline in the growth and/or reproductive success of the interacting organisms or populations (*see* **competitor**, **mutualism**, **parasite**, and **predator**).

Camptochaete arbuscula leaf **insertions**. 0.5 mm

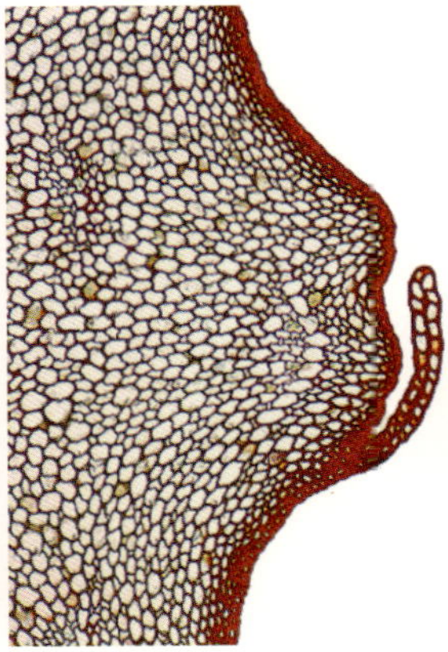

Dawsonia superba leaf **insertion** (stem xs). 100 μm

Plagiochila sp. leaf **insertion** (stem xs). 50 μm

intercalary — (1) said of branches arising below the stem tip and pushing through the **cortex** (*synonym* **endogenous**) (*the opposite of* **terminal** and **exogenous**), (2) said of growth in and around **nodes** rather than at the **apex** (•below, left).

internode — the portion of a stem between adjacent **nodes** (*the opposite of* **node**, the portion of a stem where a branch emerges or a leaf is inserted).

interrupted — said of any obvious disturbance to an arrangement that's usually highly **symmetric** (•below, right).

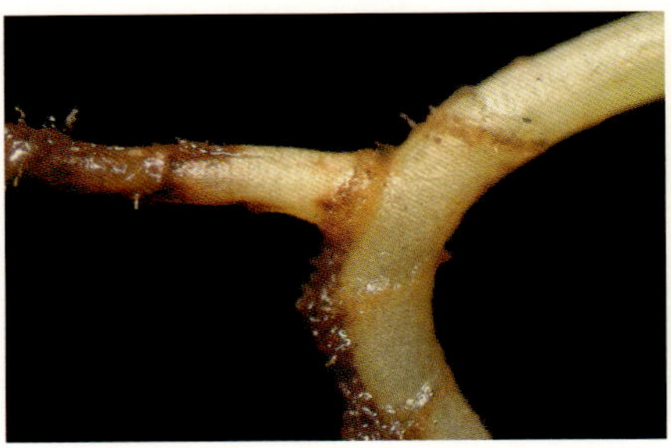

Tylimanthus saccatus **intercalary** branch with collar. 1 mm

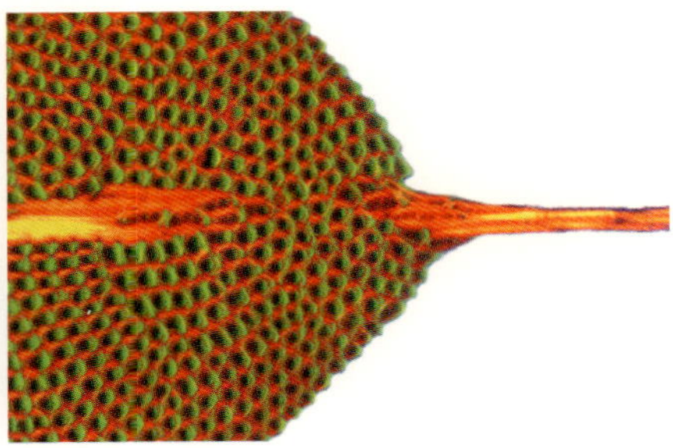

Hymenodon pilifer **interrupted** hair-point. 100 μm

Tridontium tasmanicum **intramarginal** border.
100 µm

intramarginal border — in mosses and leafy liverworts, a leaf **border** of one or more rows of specialized cells lying a short distance in from the leaf edge (•left).

intramarginal ridge — in some mosses (*Dicranum majus* among them), raised streaks toward the leaf **apex** which are **bistratose** rather than the usual **unistratose**.

intricate — interwoven or tangled.

introrse — turned inwards or towards the center (*the opposite of* **extrorse**, facing outwards or away from the center) (*compare with* **antrorse**, facing forwards or toward the tip, and **retrorse**, facing backwards or away from the tip).

inverse — said of any two things that have been turned back-to-front or head-to-tail.

involuate — inrolled, applied to leaf **margins** (*compare with* **involute**, which denotes more tightly rolled) (*the opposite of* **revolute**, rolled downwards and under at the edges).

involucral — belonging to or pertaining to the **involucre** (•below, left).

involucral bracts — modified leaves located just below a **perianth** and surrounding it, usually differing distinctly from the other leaves in their size, shape, and **margin**.

involucral scales — specialized **scales** which protect the **archegonia** of some **thallose** liverworts (•below, middle).

involucre (or **involucrum**) — a tube of **thallus** tissue which protects the **archegonia** (or rarely the **antheridia**) of some hornworts and **thallose** liverworts (•below, right) (*synonym* **perigynium**).

Marchantia sp. sterile **involucral** skirts.
1 mm

Allisonia cockaynei **involucral scales**.
1 mm

Anthoceros laminiferus horn **involucres**.
1 mm

involute — rolled inwards or upwards at the edges (•below) (*compare with* **inrolled** and **involuate**, which denote less tightly rolled) (*the opposite of* **revolute**, rolled downwards and under at the edges).

Hyophila involuta **involute** basal leaf margins. 0.5 mm

Oedicladium rufescens **involute** branch leaf. 100 µm

Polytrichum juniperinum **involute** leaf margins. 1 mm

irregular — (1) in mosses, said of **dehiscence** by decay or rupture of the **capsule** wall rather than along a line of built-in weakness such as **valves** or the rim of an **operculum** (*the opposite of* **regular**) (*synonym* **cleistocarpous**, *the opposite of* **stegocarpous**), (2) not uniform, as in the spacing of **teeth** along a leaf **margin** or the shape or arrangement of cells in a leaf blade (•below).

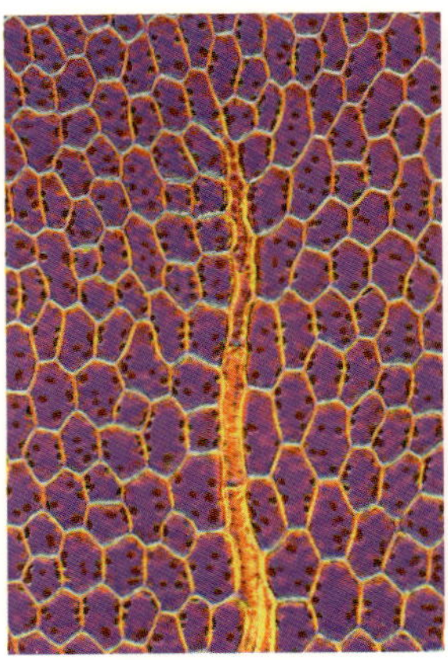

Distichophyllum microcarpum **irregular** hexagonal cells. 25 µm

Pogonatum subulatum **irregular** perigonial bract cells. 25 µm

Pendulothecium punctatum **irregular** rows of leaf cells. 25 µm

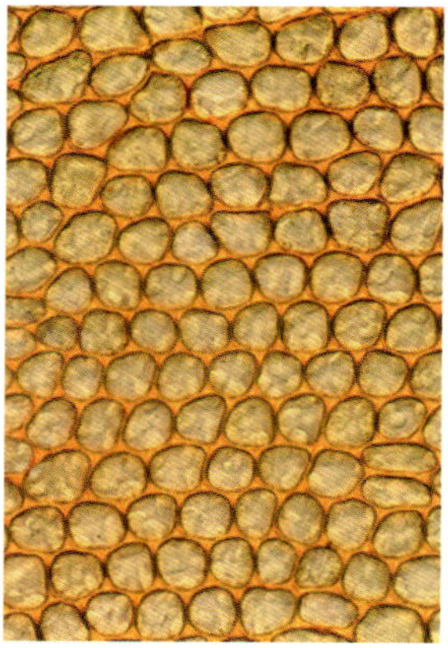

Plagiomnium maximoviczii **isodiametric** mid-leaf cells. 50 µm

isodiametric — said of cells that are about as broad as they are long, whether they're square, rounded, or polygonal (•left).

isophyllous — having only one kind of leaf on a stem, or the same type of leaf on both stems and branches (*the opposite of* **anisophyllous**).

jacket — a **sterile** wall surrounding a **capsule** or other organ. In the mosses, capsule jackets are always **multistratose**, whereas **antheridium** jackets are **unistratose**.

jacket layer — the **unistratose** wall of an **antheridium**.

jugum (plural **juga**) — a pair of leaves.

julaceous — said of a leafy **shoot** that looks like a worm or catkin because its leaves are crowded and **appressed** against the stem (•below, middle).

juniper moss — a common name given to *Polytrichum juniperinum* because its sharp bluish leaves look like miniature juniper needles.

juxtacostal — near the **midrib** (**costa**) (•below, right).

keeled — (1) sharply folded like a boat keel (•left, middle) (*synonym* **carinate**), (2) having a **wing**.

Grimmia pulvinata **keeled** leaf (xs). 50 µm

Bryum argenteum **julaceous** shoots. 1 mm

Distichophyllum microcarpum **juxtacostal** leaf cells. 100 µm

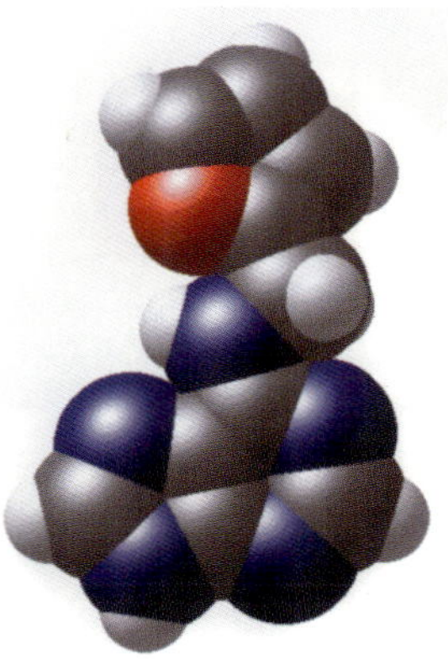

Space-filling molecular model of **kinetin**.

kinetin — one of several known **cytokinins**, a class of plant hormones that trigger cell division. A breakdown product of DNA, it's never been found in nature. Nonetheless, it's widely used in plant-growth experiments, and it's been shown to influence the growth of the **seta** in a few mosses, as well as the formation (either production or inhibition) of sex organs in *Riccia crystallina* and *Bryum argenteum*. However, it has no effect at all in some other bryophytes (•left).

kinetochore — a specialized portion of a **chromosome** where a spindle fiber attaches.

knobbed — swollen at the end (•below, left).
lacerate — deeply and irregularly cut or torn.
laciniate — (1) finely and deeply dissected, (2) fringed with **cilia** (•below, middle).
lacunose — (1) said of cells that have irregular or seemingly empty **lumina** because their walls are deeply **porose** or thick (•below, right), (2) having deep depressions or holes.
laevigate — polished.

Frullania subhampeana **knobbed** elaters. 50 µm

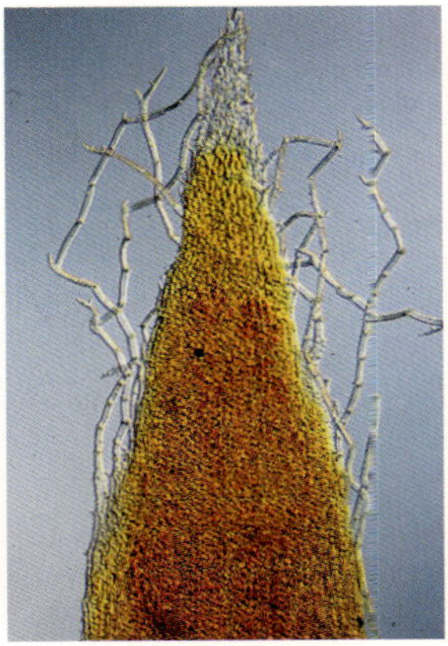

Hedwigia ciliata **laciniate** perichaetial leaves. 100 µm

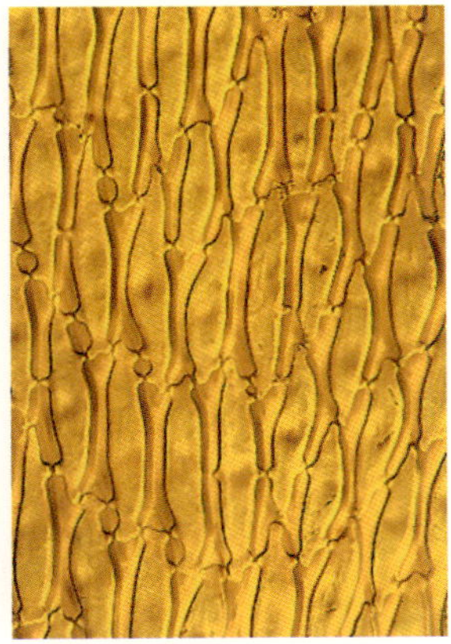

Pseudospiridentopsis horrida **lacunose** leaf cells. 25 µm

lamella (plural **lamellae**) — (1) a wall-like rib or flap running lengthwise down the leaves of some mosses and the thalli or **perianths** of some liverworts (•below and next page), (2) a projection on the back of a **peristome tooth** (**trabecula**).

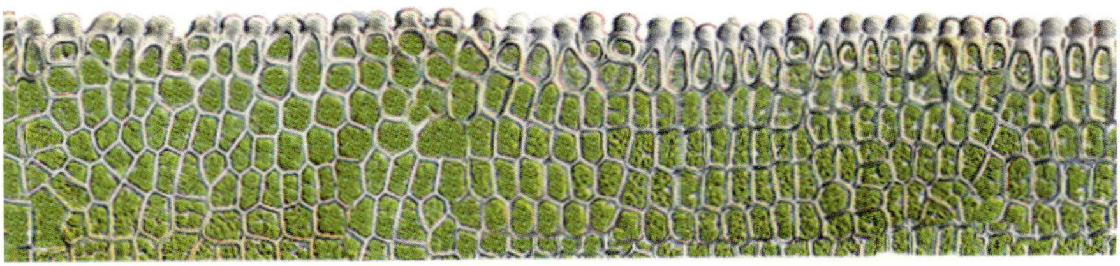

Polytrichum juniperinum **lamella** (wm). 50 µm

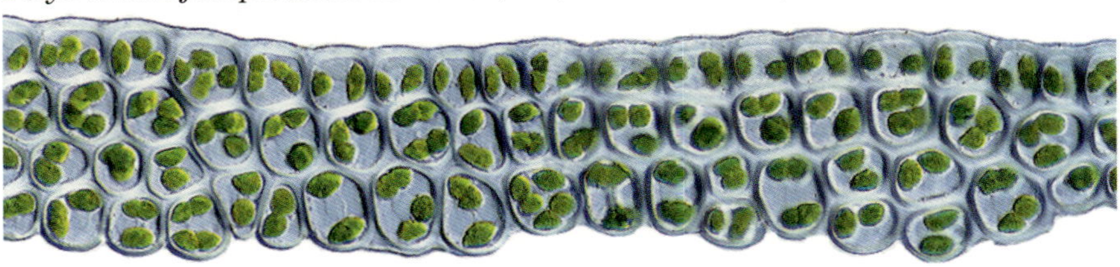

Pogonatum subulatum **lamella** (wm). 50 µm

lamella (continued)

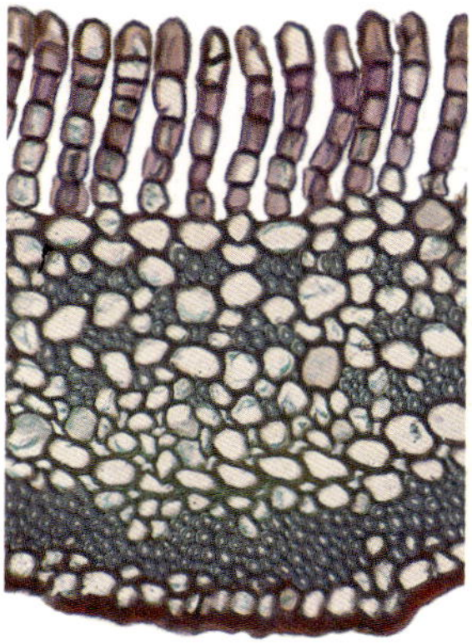

Dawsonia superba **lamellae** (xs).
50 µm

Polytrichadelphus magellanicus **lamellae** (xs).
100 µm

Atrichum angustatum **lamella** (wm).
50 µm

lamellate — having **lamellae**.

lamina (plural **laminae**) — the flat blade of a leaf or **thallus** (usually one or a few cells thick), not including the **midrib** (**nerve** or **costa**) if present.

lanate — woolly (*see* **hairy** for a comparison of similar terms).

lanceolate — lance-shaped, narrow and **tapering** from the base (•below) (*compare with* **ovate**, egg-shaped).

Lanceolate leaves (from left): *Leptodictyum riparium, Kindbergia praelonga,* and *Racopilum robustum* (various magnifications).

lanuginose — woolly, **tomentose** (•below, left).

lateral — at the side or at 90° to the stem (•below, middle and right) (*the opposite of* **terminal**).

lateral-intercalary — (1) in leafy liverworts, said of **intercalary** branches that are produced in the **axils** of **lateral leaves**, (2) any lateral branch located well below the **apex**.

Racomitrium lanuginosum **lanuginose** moss. 5 mm

Pyrrhobryum mnioides **lateral** inflorescences. 1 mm

Balantiopsis diplophylla **laterally** arranged cell rows. 50 µm

lateral leaves — the larger leaves of leafy liverworts, arranged in two opposing rows along the sides of the stem (•below, right) (*compare with* **underleaf**, mostly in one row, smaller, and shaped differently).

Laurasia — an ancient supercontinent that broke up to form present-day North America and much of Eurasia.

lax — (1) said of cells that are thin-walled and bulging when **turgid**, but tend to collapse when they dry out (•below, left), (2) said of parts that are somewhat distant from each other.

Rhaphidostichum boschii ssp. *thelidictyon* **lax** alar cells of leaf base. 100 µm

Chiloscyphus fissistipus **lateral** leaves and underleaves. 1 mm

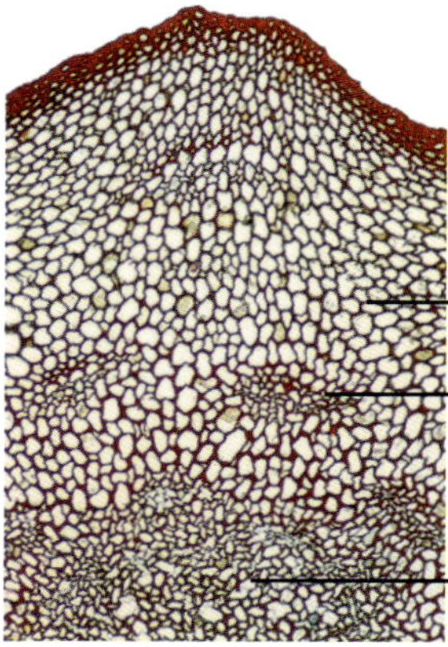

Dawsonia superba **leaf traces** (stem xs).
100 μm

leaf trace — in mosses, an extension of the **midrib** of a leaf into the **cortex** of the stem. A true leaf trace connects with conducting cells in the **central strand**, whereas a false leaf trace ends blindly in the **cortex** (•left).

leafy — (1) said of a **gametophyte** made up of regularly arranged leaves and stems (•below) (*the opposite of* **thallose**), (2) said of some **thallose** liverworts such as *Treubia* and *Fossombronia* which have leaf-like **wings**.

Heteroscyphus cymbaliferus **leafy** gametophyte.
1 mm

Schistochila sp. **leafy** shoot.
1 cm

Lejeunea sp. **leafy** shoot.
1 mm

leg. — an abbreviation for **legit**., a Latin word meaning *collected by*, often noted on herbarium collections.

***Lejeunea*-type branching** — in leafy liverworts, a type of branching in which a branch grows from a **cortical** cell and has a **basal** collar or **sheath** that develops from an adjacent leaf base (for a comparison of other types of branching, *see* ***Acromastigum*-type branching**).

lenticular (or **lentiform**) — shaped like a lentil or a double-**convex** lens (•below, left).

lentigenose — appearing dusty.

leptodermous (or **leptodermatous**) — (1) thin-skinned, (2) thin-walled (•below, middle and right) (*the opposite of* **pachydermous**, thick-walled).

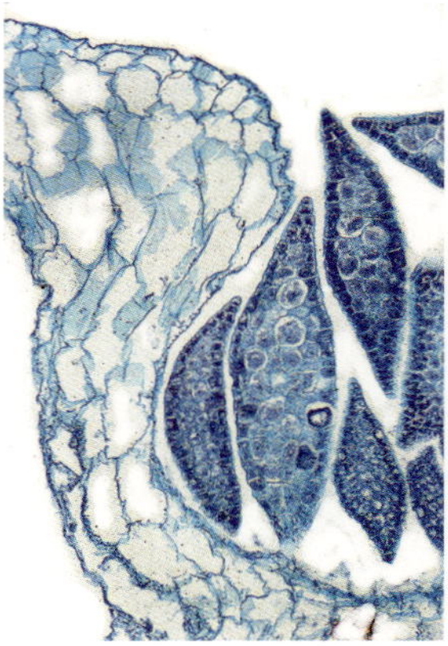

Marchantia sp. **lenticular** gemmae (ls). 100 μm

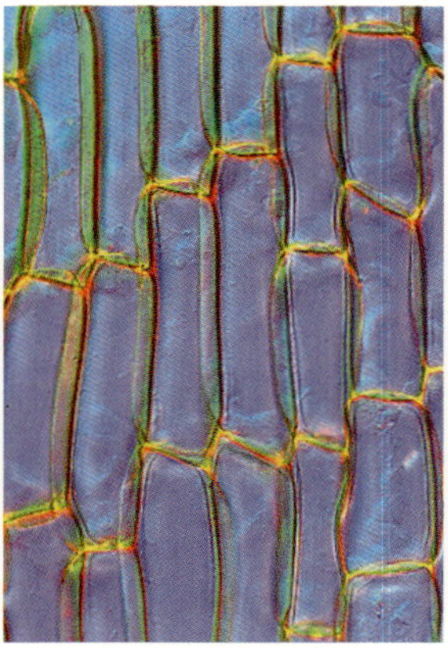

Funaria hygrometrica **leptodermous** cells. 50 μm

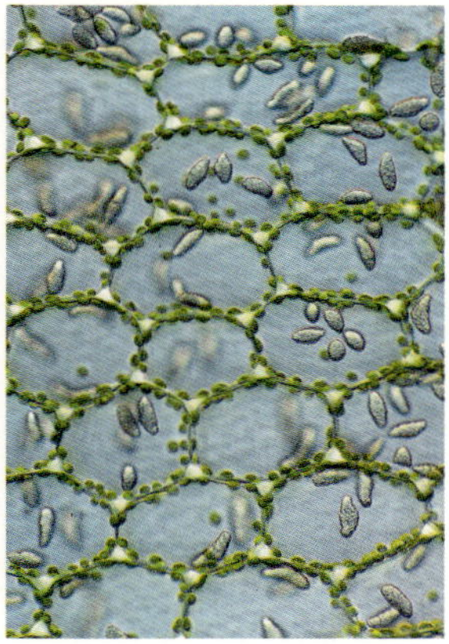

Jungermannia inundata **leptodermous** cells. 50 μm

leptoids — conducting cells in the **central strand** and **setae** of some mosses (•below, left).

leptome — collective term for all the **leptoids** and **parenchymatous** cells in the **central strand** of some mosses.

leucocyst — a large, empty **hyaline** cell in the leaves of species of *Sphagnum* and the moss family Leucobryaceae (•below, right) (*synonym* **hyalocyst**).

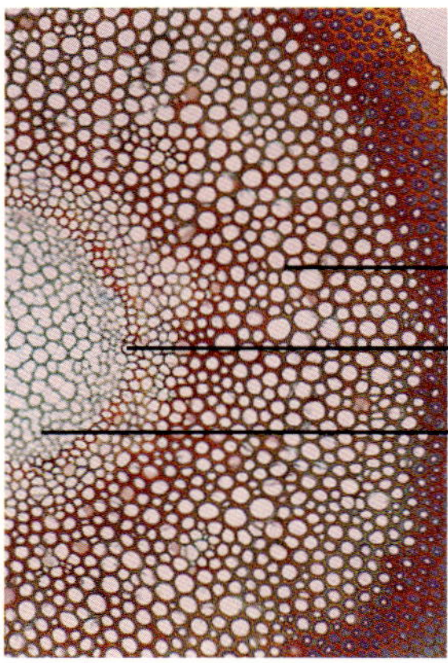

Dendroligotrichum dendroides **leptoids**. 100 μm

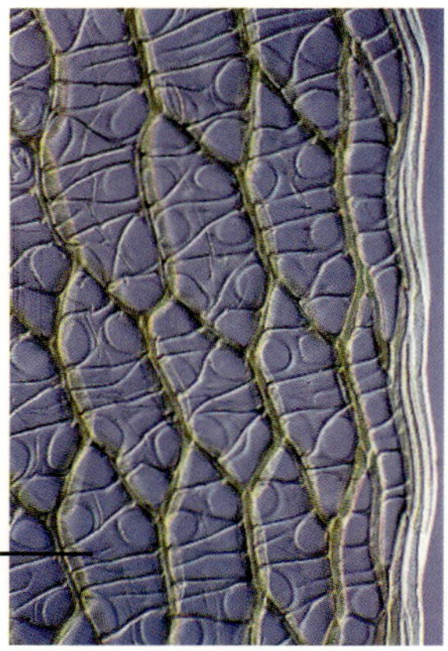

Sphagnum squarrosum **leucocysts**. 25 μm

lid — **operculum**, the cover of a moss **capsule**.
lignicolous — growing on lignum (wood that's lost its bark in the first stages of rotting) (•below, left).
lignose (or **ligneus**) — resembling wood in texture.
ligulate — strap-shaped (*synonym* **lorate**) (*compare with* **lingulate**, tongue-shaped, therefore wider).
limb — the upper part of a leaf (*the opposite of* **base**).
limbate — **bordered** (•below, middle and right).
limbidium (plural **limbidia**) — **border** (•below, middle and right).

Chiloscyphus echinellus **lignicolous**.
1 mm

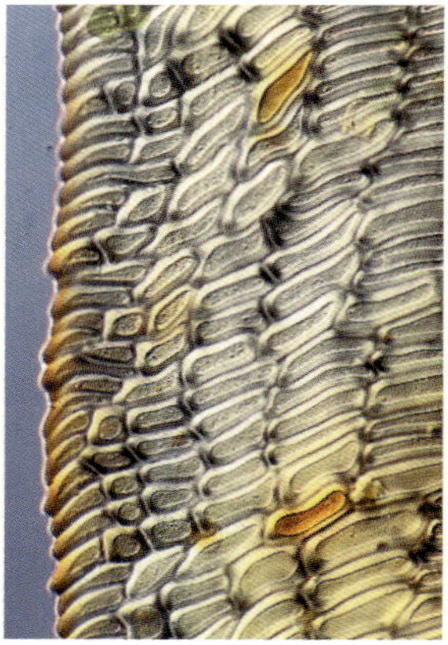

Polytrichum juniperinum hyaline **limbidium** cells.
100 μm

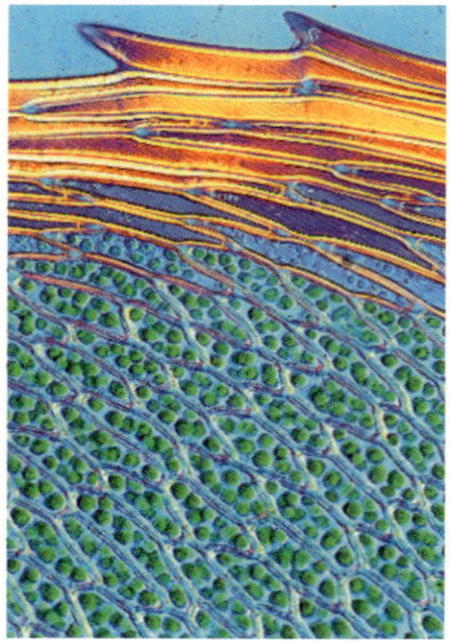

Bryum billardierei **leaf limbidium**.
100 μm

limnophilous — thriving in fresh water.
lineage — the line of evolutionary descent of a group of organisms.
lineage pattern — the line of descent of a tissue or an organ from an **initial** cell or a **primordium** to maturity.
linear — long and narrow with nearly parallel edges (•below and next page).

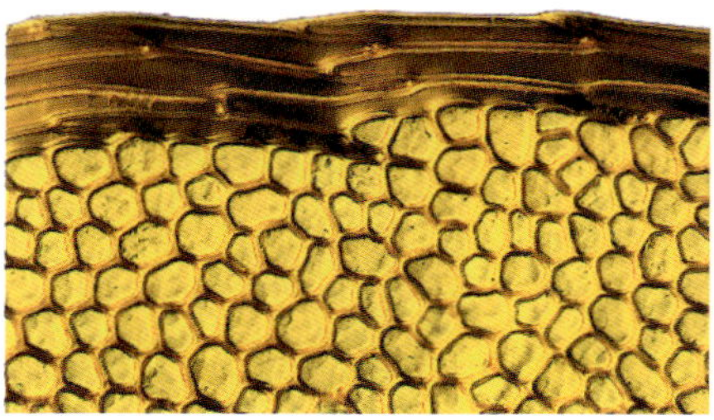

Plagiomnium cuspidatum **linear** leaf border cells.
100 μm

Polytrichum juniperinum **linear** leaf sheath cells.
50 μm

linear (continued)

Entodon seductrix **linear** mid-leaf cells.
25 µm

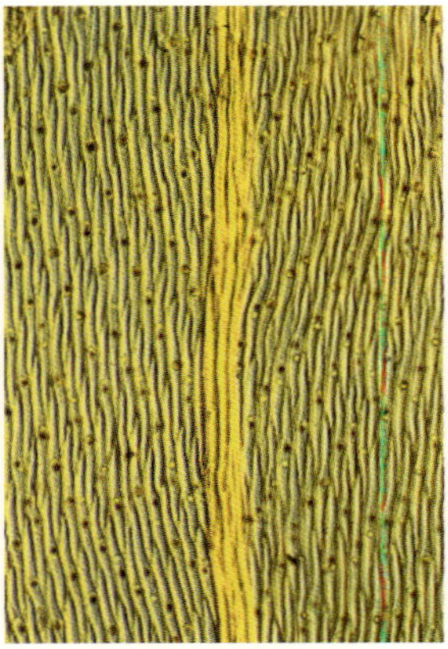

Pseudobarbella laosiensis **linear** leaf cells.
100 µm

Spiridens veillardii **linear** leaf sheath cells.
50 µm

lineate — lined, marked with thin parallel lines (*synonym* **striate**).

lingulate (or **linguiform**) — tongue-shaped (•below, left) (*compare with* **ligulate**, strap-shaped, which is narrower).

lipophilic — readily dissolving in fatty substances (*the opposite of* **lipophobic**).

lipophobic — not dissolving in fatty substances (*the opposite of* **lipophilic**, readily dissolving in fatty substances).

lobate (or **lobed**) — divided into **lobes** (•below, middle and right).

Neckeropsis calcicola **lingulate** stem leaf.
0.5 mm

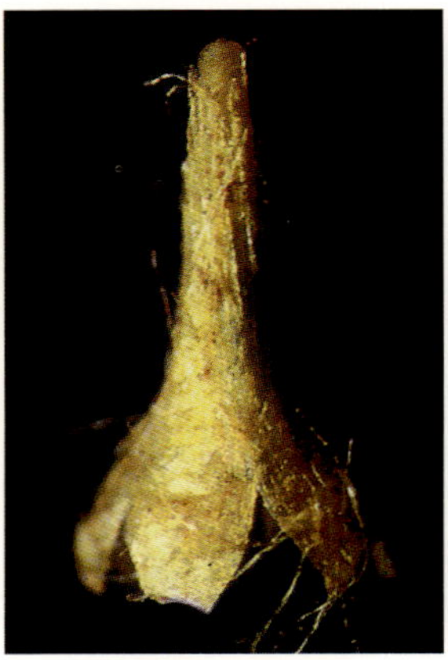

Racopilum robustum **lobate** calyptra.
1 mm

Racomitrium crispulum **lobate** calyptra.
1 mm

lobe — a major division of a leaf or other plant organ (•below).

Dicnemon dixonianum calyptra **lobes**. 1 mm

Neohodgsonia mirabilis thallus **lobes**. 1 mm

Catharomnion ciliatum calyptra **lobes**. 1 mm

lobule — (1) in a **complicate-bilobed** leaf, the smaller **lobe**, (2) a sac formed by an **inrolled** rear leaf **margin**, as in leafy liverworts of the family Lejeuneaceae and the genus *Radula* (•below), (3) any small **lobe**.

lochmocolous — growing in thickets or copses.

longevity — (1) long life, (2) the length of life.

longitudinal — placed or arranged lengthwise.

lorate — narrow with parallel **margins** (*synonym* **ligulate**, strap-shaped).

loricate — (1) armored with hard **scales**, (2) equally narrow throughout.

Radula buccinifera **lobules**. 1 mm

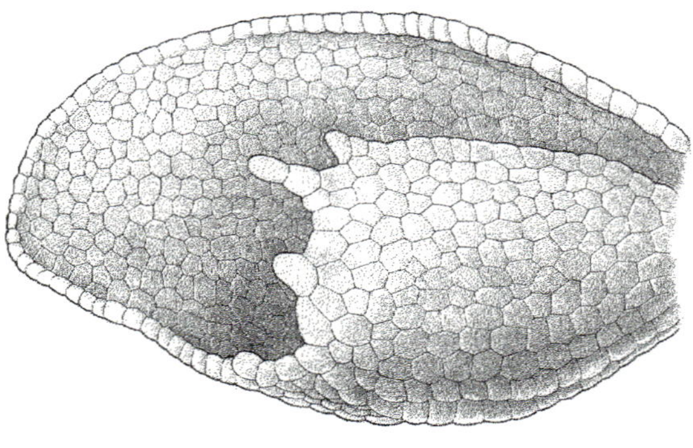

Austrolejeunea olgae **lobule**. 50 mm

lumen (plural **lumina**) — the cavity of a cell inside the cell walls (•below).

Grimmia pulvinata dumb-bell-shaped leaf **lumina**. 50 µm

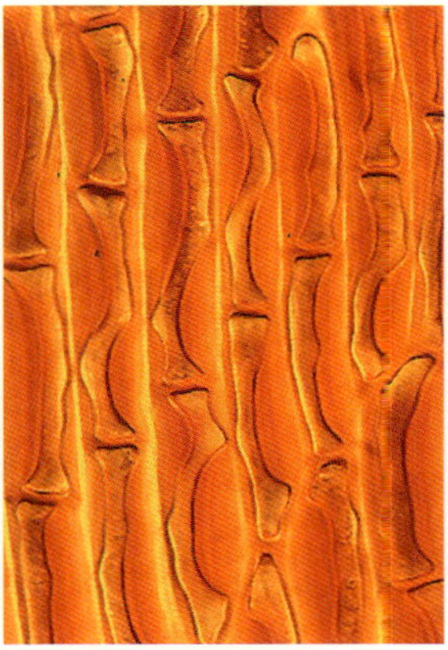

Macromitrium longipes sinuose leaf cell **lumina**. 25 µm

Schistidium apocarpum leaf cell **lumina**. 25 µm

Acromastigum anisostomum leaf cell **lumina**. 100 µm

Isotachis montana leaf cell **lumina**. 50 µm

Metzgeria sp. leaf cell **lumina**. 100 µm

luminal — part of the **lumen** or projecting into it.

luminous mosses — a common name for *Mittenia*, *Schistostega*, and other mosses which live in deep shade and appear to glow. The word *luminous* is misleading, because the mosses don't actually *produce* light as a firefly or glow-worm does—what looks like a glow is merely light that's *reflecting* from the **inflated** cells of their **persistent protonemata**. Those inflated cells gather and focus light onto their **chloroplasts**, a trait that allows the mosses to survive in the ultra-low light of their **habitats**—caves, deep cracks, and overhangs.

lunate (or **lunulate**) — shaped like a crescent moon (•below, left).
lurid — wan, ghastly pale, or dingy yellow-brown.

Lunularia cruciata **lunate** gemma cup.
1 mm

Weymouthia cochlearifolia **lustrous** leaves.
10 mm

Rhytidiadelphus squarrosus **lustrous** leaves.
1 mm

lustrous — having a sheen or gloss that readily reflects light (•above, middle and right) (*synonym* **nitidous**).
luteous (or **luteus**) — bright yellow (•below, left).
luticolous — growing in mud, clay, or mires.
lux — a unit of illumination.
lyrate — shaped like a lyre.
Macaronesia — an area of the North Atlantic Ocean which includes the Azores, Madeira, the Canaries, and the Cape Verde Islands.
macrocladous — producing large or long branches.
macronemata (singular **macronema**) — in mosses, large and freely branched **rhizoids** restricted to leaf **axils** and branch **insertions** (•below, right) (*compare with* **micronemata**, which are short and only sparingly branched, and arise more randomly on stems or leaf bases).

Phaeoceros hirticalyx **luteous** spores.
50 μm

Campylopus clavatus **macronemata**.
1 mm

Bartramia pomiformis var. *elongata* **mammillae**.
100 µm

mammilla (or **mamilla**) (plural **mammillae** or **mamillae**) — (1) a low, rounded projection on a moss **operculum**, (2) on the surface of a cell, a bulge with a nipple-like tip. A mammilla is a bulge in a uniformly thickened cell wall, whereas a **papilla** is a local thickening of the cell wall. The difference can be seen only by sectioning the cell (•left) (*compare with* **scindula**, the protruding end of a **prosenchymatous** cell).

mammillate (or **mamillate**) — **mammillose**.

mammillose (or **mamillose**) — bulging with a blunt central projection (•left) (*synonym* **mammillate**).

marcescent — withered, dry, or dead but not yet fallen off.

margin — (1) a leaf **border**, (2) the edge of a structure such as a leaf (•below).

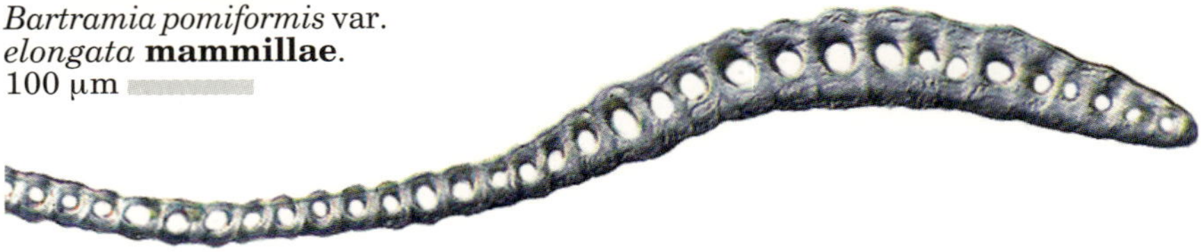

Ptychomnion aciculare leaf **margin** (cross-section). 50 µm

marginal — at the edge or referring to the edge of a structure such as a leaf (•below, left and middle).

marsupium (plural **marsupia**) — a swollen and elongate pouch of stem tissue that encloses and protects the **sporophyte** in some jungermannialean liverworts (•below, right). A highly elongated marsupium is sometimes called a **marsupidium**.

Pseudobryum speciosum **marginal** spines.
100 µm

Metzgeria sp. **marginal** hairs.
1 mm

Tylimanthus saccatus **marsupium**.
1 cm

mat — a growth form in which stems are flattened on the **substratum** and densely interwoven (•below, left and middle).

m-chromosomes — minute supernumerary **chromosomes** found in the nuclei of some bryophytes. Unlike typical chromosomes, they often remain condensed during interphase, and they lack **chiasmata**. During **meiosis** they can appear to be univalents because they separate and move to the poles earlier than other chromosomes do. They're denoted in chromosome counts by a + symbol, for example n = 11+1.

median — middle or central.

median leaf cells — in the leaves of mosses and leafy liverworts, the cells located about 2/3 up from the base.

medulla — the central region of a stem, surrounded by the **cortex** (•below, right).

medullose — pithy or spongy.

Thuidium furfurosum **mat** growth habit.
10 mm

Wijkia extenuata **mat** growth habit.
10 mm

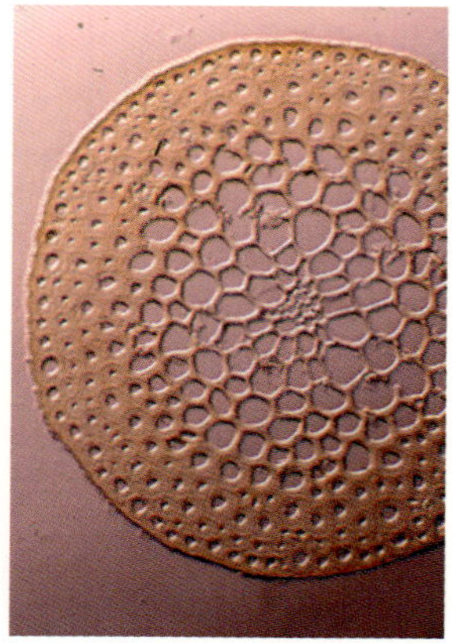

Leptostomum inclinans **medulla** and cortex (xs).
100 μm

meiosis — a type of **chromosomal** division by which the number of sets of chromosomes is reduced from the **diploid** (2n) state to the **haploid** (1n) state. It involves replication of the chromosomes followed by two divisions (called meiosis I and II), and produces four daughter nuclei, each containing one complete set of chromosomes. During meiosis, corresponding segments of **chromatids** from **homologous** chromosomes are swapped, a process called crossing-over. As a result, the progeny and parent genotypes differ. In **mitosis**, in contrast, the progeny and parent genotypes are identical.

membranaceous — film-like, thin and transparent.

membrane gaps — irregular openings in the walls of **hyaline** cells (**leucocysts**) of the stem leaves in species of *Sphagnum*.

Hypopterygium rotulatum **micronemata.** 1 mm

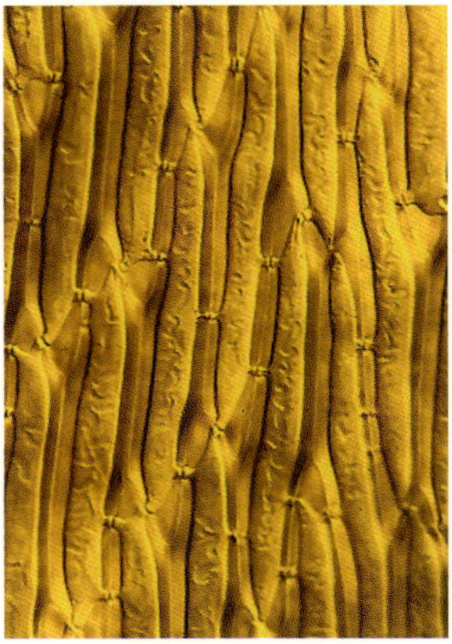

Eucamptodon inflatus **middle lamellae** of leaf cells. 25 µm

meristem — at the base of hornwort **sporophytes**, a zone of continual cell division and growth.

merophyte — a **primordial** segment of a **shoot** that's produced by one of the **cutting faces** of the shoot's **apical** cell. The merophyte in turn gives rise to stem and leaf tissues.

mesic — moist, intermediate between very wet and very dry (*compare with* **hydric** and **xeric**).

mesomorphic — structurally adapted for surviving in moist **habitats** (*compare with* **hydromorphic**, **hygromorphic**, and **xeromorphic**).

mesophilous — preferring moist **habitats** (*compare with* **hydrophilous hygrophilous**, and **xerophilous**).

mesophyte — a plant adapted to moist or moderately humid **habitats** (*compare with* **hydrophyte**, **hygrophyte**, and **xerophyte**).

***Microlepidozia*-type branching** — in leafy liverworts, a type of branching in which a branch replaces the **antical** half of a leaf (for a comparison of other types of branching, *see* ***Acromastigum*-type branching**).

micrometer — (1) (or **micrometre**, symbolized by **µm**) one-millionth of a meter or one-thousandth of a millimeter (formerly called a **micron** and symbolized by µ), (2) a mechanical or optical device for measuring microscopic distances.

micronemata (singular **micronema**) — short and slender, sparingly branched **rhizoids** arising here and there on stems or leaf bases (•top left) (*compare with* **macronemata**, which are large and freely branched, and are restricted to leaf **axils** and branch **insertions**).

microphyllous branch — a branch with leaves distinctly smaller than usual (•below) (*similar to* **flagelliform branch**).

middle lamella — the layer that cements together the primary walls of adjacent cells (•left).

mid-leaf cells — leaf cells in an area that's halfway between the base and the tip, and in mosses also halfway between the **margin** and the **nerve**.

Acromastigum anisostomum **microphyllous branches.** 1 cm

Radula multiamentula **microphyllous** branches. 1 mm

midrib — in some mosses and liverworts, a vein in the center of a leaf or **thallus** (•below, left and middle) (*synonyms* **nerve** and **costa**).

minerotrophic — said of anything that's rich in mineral nutrients, for example water running off or draining a soil that has a high percentage of clay particles.

Metzgeria sp. **midrib**.
100 µm

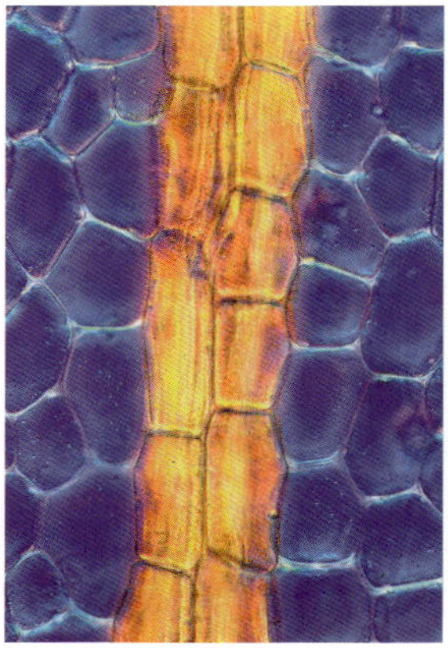

Austrometzgeria saccata **midrib**.
50 µm

Macromitrium longipes **mitrate** calyptrae.
1 mm

mitosis — a type of **chromosomal** division in which the genotype of the two new daughter cells is identical to that of the parent cell (*compare with* **meiosis**, in which the daughter genotypes are not identical to that of the parent cell).

mitrate (or **mitriform**) — shaped like a bishop's cap (miter), **conical** with an **entire** or regularly **lobed** base (•above, right) (*compare with* **cucullate**, conical and split down one side).

mixohydric (or **mesohydric**) — conducting water both externally and internally (*compare with* **endohydric**, conducting water internally, and **exohydric**, conducting water externally).

monandrous — bearing only one **antheridium**.

moniliform — resembling a string of beads (•left).

monoicous — producing male and female sex organs (**antheridia** and **archegonia**) on the same plant (*the opposite of* **dioicous**) (*not the same as* **monoecious**, which applies to the **diploid sporophytes** of vascular plants). If the male and female are in separate **inflorescences**, the arrangement is termed **autoicous**. If they're in the same inflorescence but each sex is grouped or clustered off by itself, that's called **paroicous**, whereas if they're in the same inflorescence but *not* grouped or clustered, then that's termed **synoicous** or **gynandrous**.

monolete spore — a **spore** that has a single scar on its surface (*compare with* **trilete spore**).

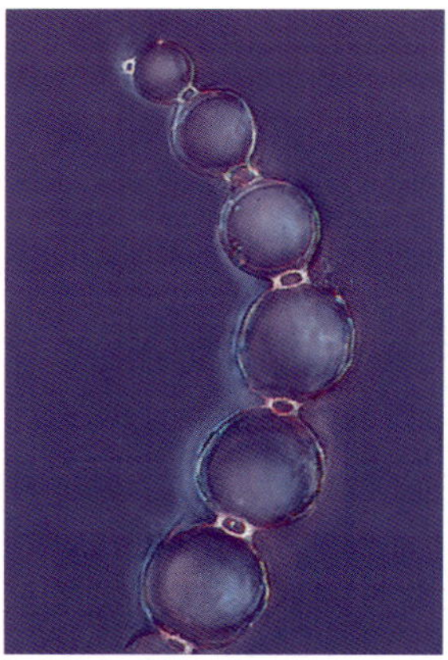

Sphagnum cristatum **moniliform** leaf cells (xs).
50 µm

monomorphic (or **monomorphous**) — having only a single form.

monopodial branching — a pattern of branching with numerous secondary shoots or stems coming off a main stem. Monopodial growth is unlimited because the **inflorescences** are all **lateral** and therefore can't halt the growth of the main stem.

monostromatic — one cell layer thick (*synonym* **unistratose**).

monoterpenoid — a **terpenoid** that's made up of two isoprene units (*see* **terpenoids**).

monotypic — (1) said of a **genus** that contains only one **species**, (2) said of any taxonomic rank that contains only one member in the rank below it, for example a family with only one genus.

montane — referring to mountains.

morphology — outward form or appearance.

moss balls — moss **cushions** that have broken loose and have been rounded into spheres by constant rolling in the wind. They can form any place that's windy, open, and flat and where mosses are dislodged now and then by freeze-thaw cycles, violent winds, or pawing animals. The most spectacular moss balls are the whimsically named glacier mice of Alaska and Iceland. Looking like mice scurrying over the ice, they're chunks of *Drepanocladus revolvens* and *D. berggrenii*. In the extensive sand dunes of Alaska's Amchitka Island, fragments of the rock-dwelling *Schistidium apocarpum* (•left) that are loosened by the wind round up into miniature tumbleweeds as they bounce over the surface of the sand. Moss balls can form even where there's little wind—loose chunks of *Leucobryum* cushions scratched apart by birds or pawing animals can survive as drifting vagrant "polsters".

Schistidium apocarpum mature capsules.
1 mm

mucilage — a clear, gel-like, slimy secretion of some liverworts (•below, left) (*synonym* **slime**).

mucilage cavity — a cavity in a **thallus** containing **mucilage** and often the cyanobacterium *Nostoc.*

mucilage cell — a **mucilage**-containing cell near an **apex** or sex organ, or in a **thallus** cavity.

mucilage papilla — a cell that secretes **mucilage** (•below, right) (*synonym* **slime papilla**).

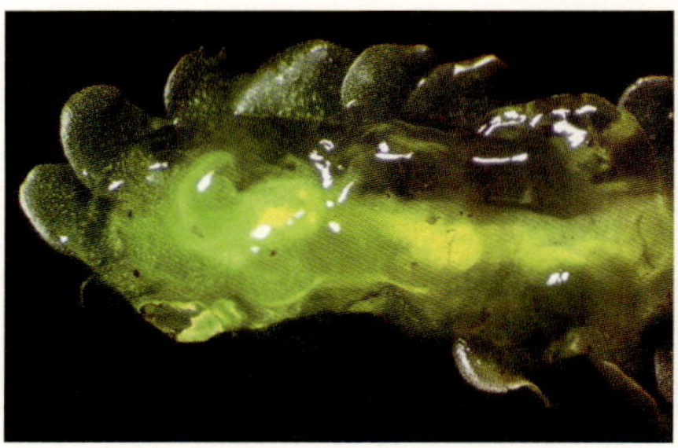

Treubia lacunosa shoot underside with **mucilage**. 5 mm

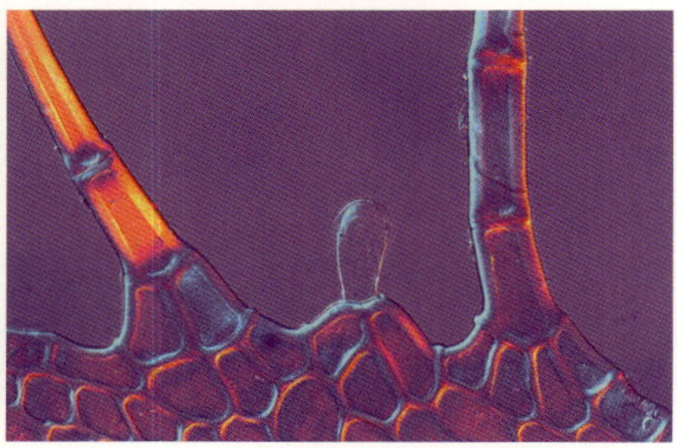

Schistochila nobilis **mucilage papilla**. 50 μm

mucose (or **mucous**) — covered with **slime**.
mucro (plural **mucrones**) — a short, **abrupt** point (•below, left).
mucronate — ending **abruptly** in a short point (**mucro**) (•below, left) (*compare with* **apiculate**, ending in a point somewhat longer than a **mucro**, and **cuspidate**, ending in an even longer point that's also stouter).
mucronulate — ending in a minute short point.
multi- — a prefix meaning *many*.
multicellular — made up of many cells (•below except for top left and bottom right).
multicostate — in some mosses and liverworts, having several **costae** (•below, bottom right).

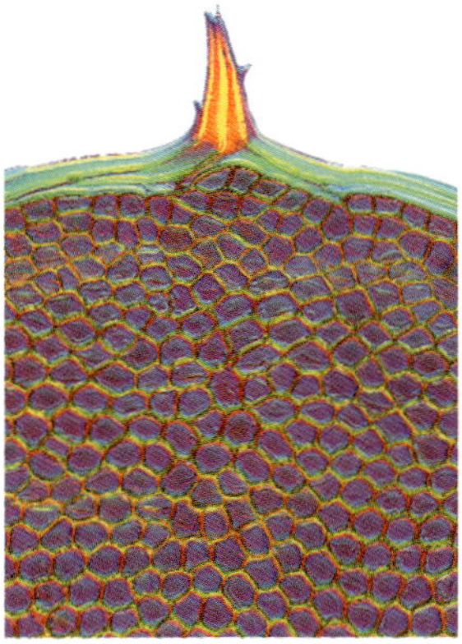
Distichophyllum pulchellum **mucro**.
100 µm

Cladomnion ericoides **multicellular** spore.
50 µm

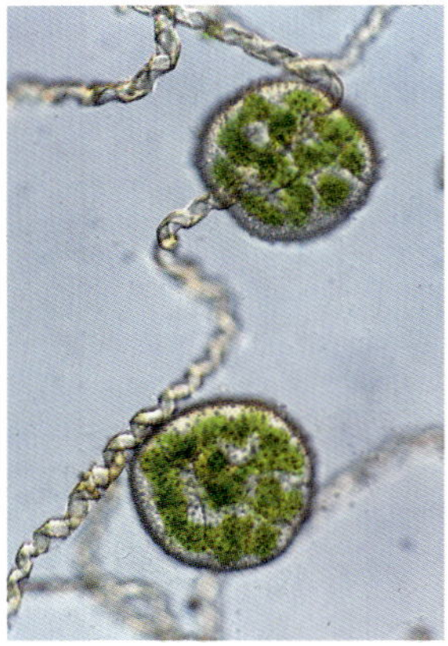
Dendroceros validus **multicellular** spores.
100 µm

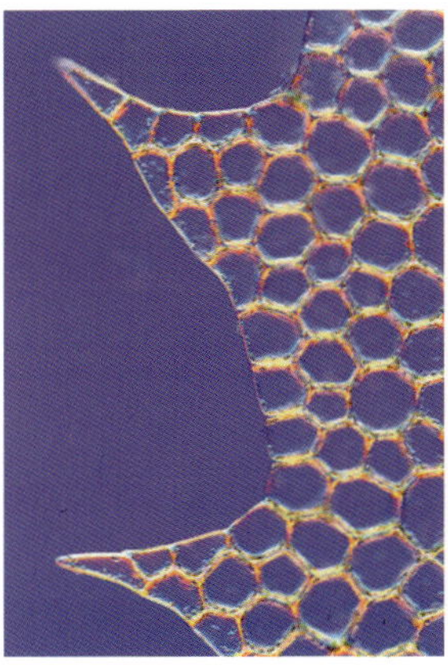
Achrophyllum dentatum **multicellular** leaf teeth.
50 µm

Marsupidium setulosum **multicellular** projections. 1 mm

Antitrichia curtipendula **multicostate** leaf base.
100 µm

multifid — repeatedly divided.
multipapillose — with many **papillae** per cell (•below).

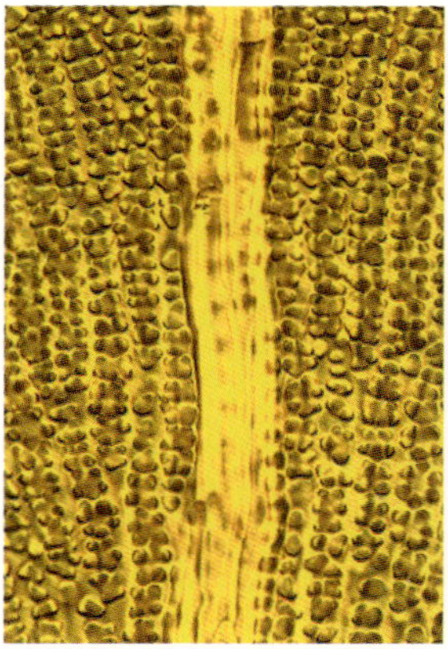

Floribundaria aurea ssp. *nipponica* **multipapillose** cells. 25 μm

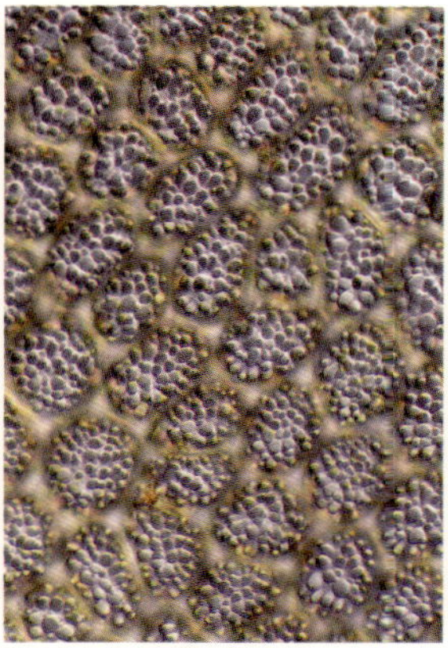

Saccogynidium australe **multipapillose** cells. 100 μm

Acrobolbus cinerascens **multipapillose** cells. 100 μm

Schistochila muricata **muricate** leaf xs. 100 μm

multipinnate — several times **pinnately** branched (*compare with* **pinnate**, once-pinnately branched, and **bipinnate**, twice-pinnately branched).
multiseriate — composed of many rows of cells.
multistratose — having many layers (•below, right).
muricate — roughened with many sharp points or spines (•left and below, left).
muriculate — roughened with numerous minute sharp points.
muticous — in mosses, said of a leaf that lacks an **awn** or **hair-point** but ordinarily would have one.
mutualism — an interaction which enhances the growth and/or reproduction of two organisms or populations living on or near each other.
mycorrhizal — harboring fungal hyphae within or among the cells of a plant.

Schistochila muricata **muricate** leaf surface. 100 μm

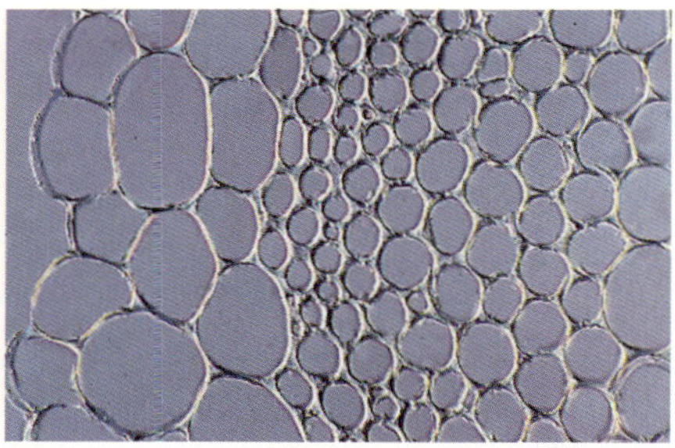

Sphagnum cristatum **multistratose** hyaloderm cortex. 100 μm

Leptostomum macrocarpon archegonia **necks**.
100 µm

n — the **haploid chromosome** number of an organism or tissue, for example n = 4. If **m-chromosomes** are present as well, they're denoted by the symbol +, for example n = 11+1 (*compare with* **x**, the basic haploid chromosome number of a **species** or other **taxon**).

naked — smooth, lacking hairs (•below, left) (*similar to* **glabrous**).

nannandrous (or **nanandrous**) — sexually **dimorphic**, with miniature male **gametophytes** borne on normal-sized female plants (*see also* **dwarf male**, **phyllodioicous**, and **pseudautoicous**).

nanogranules — very small particles.

navicular — boat-shaped (•below, middle) (*synonym* **cymbiform**).

neck — (1) in mosses, the **sterile** portion of a **capsule** between the base of the **theca** and the top of the **seta** (•below, right) (*synonyms* **apophysis** and **hypophysis**), (2) the cylindrical upper portion of an **archegonium** (•left).

neck canal cells — the cells that temporarily fill the center of the **neck** of an immature **archegonium** (*synonym* **canal cells**).

Calliergonella cuspidata **naked** stem apices.
1 mm

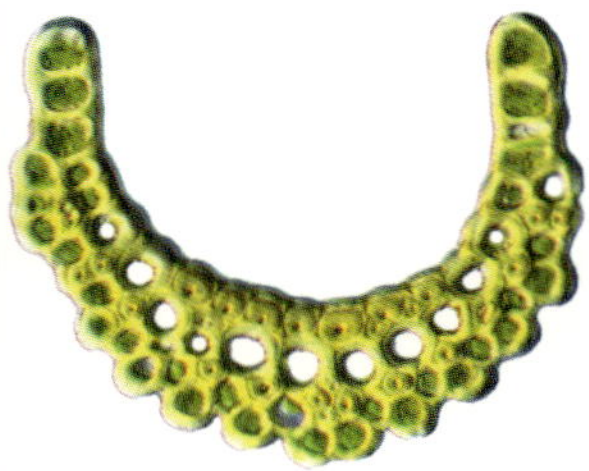

Campylopus clavatus **navicular** leaf section.
50 µm

Bryum billardierei sterile capsule **neck**.
1 mm

nematodontous — said of a moss **peristome** which is made up of whole (dead) cells with evenly thickened walls. Diverse families of mosses have such peristomes (*compare with* **arthrodontous** peristomes, which are made up of only *parts* of dead cells).

nematogon — a **rhizoid initial**.

nematogonous — (1) in mosses, said of **gemmae** which are formed directly on **protonemata**, (2) thread-like.

neotropic — a pattern of distribution restricted to the American tropics.

nerve (adjective **nervate or nervose**) — the **midrib** of a leaf (*synonyms* **vein** and **costa**).
niche — the role of an organism in its ecosystem.
nidulant — (1) lying in a cup or cavity, (2) nested on a **protonematal mat**.
nitidous — shiny or glossy, sometimes with an opalescent or metallic sheen (•below, left and middle).

Pseudotaxiphyllum falcifolium **nitidous** leaves. 1 mm

Pohlia cruda **nitidous** leaves. 1 mm

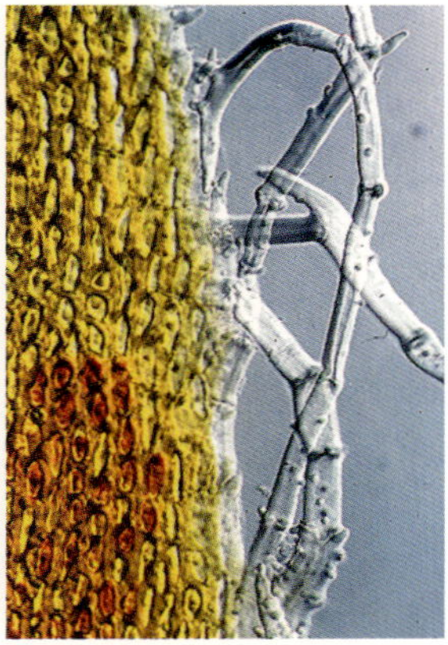

Hedwigia ciliata **nodose** cilia of inner perichaetial leaves. 50 μm

nitrophilous — thriving on **substrata** rich in nitrogenous compounds.
node — the portion of a stem where a branch emerges or a leaf is inserted (*the opposite of* **internode**).
nodose — **knobbed** (•above, right).
nodulose — minutely **knobbed** (•below).

Racomitrium crispulum **nodulose** lower leaf cells. 10 μm

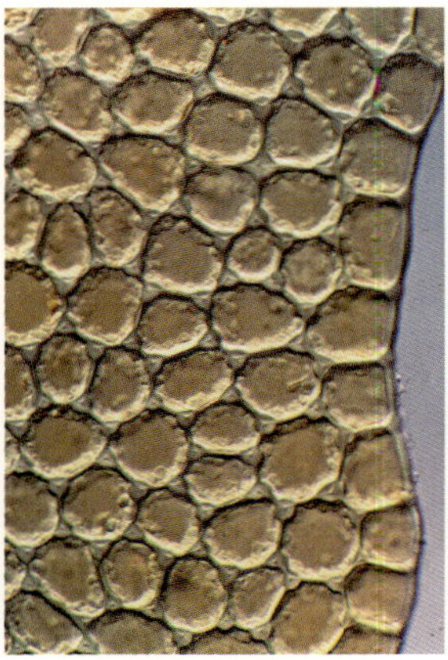

Andrewsianthus confusus **nodulose** trigones. 100 μm

Jubulopsis novae-zelandiae **nodulose** cell walls. 100 μm

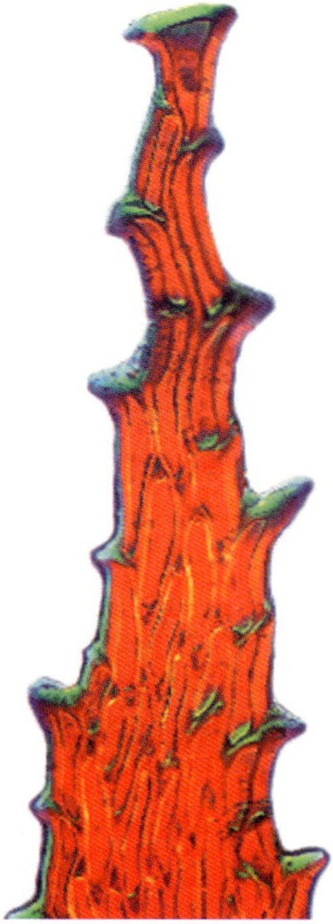

Antitrichia curtipendula **ob**triangular terminal cell of acumen.
50 μm

nondisjunction — (1) meiotic: the failure of **homologous chromosomes** to separate during **meiosis**, (2) mitotic: the failure of sister **chromatids** to separate during **mitosis** (*see* **autopolyploidy**).
normal — typical, unexceptional.
nullinervate — in mosses, lacking a **midrib** (**costa**) (*synonyms* **ecostate** and **enervate**).
nunatak refugia — scattered areas that somehow escaped being glaciated during Pleistocene times and so were reservoirs of species, including bryophytes, for re-colonizing surrounding areas quickly as soon as the glaciers melted away.
nurse cells — a layer of nutritive cells in a developing **sporangium** (*synonym* **tapetum**).
nurse logs — felled logs that are deliberately left behind by loggers during forest clear-cutting to encourage forest regeneration. The logs are intended as habitat for **mat**-forming bryophytes, especially mosses, which in turn will be invaded by the seedlings of valuable forest trees and thus act as a forest "nursery". The seeds of worthless "junk" trees can not germinate in the mats, and so nurse logs speed up regeneration of the forest and shorten the "turn-around" time to the next clear-cutting.
nutant — nodding or drooping more than 90° from vertical (•below, left and middle).
ob- — a prefix meaning *inverted*, *reversed*, or *the other way around*, as in **obconical**, **oblate**, and obtriangular (•left).
obconical — having an upside-down cone shape.
oblate — wider than long (•below, right) (*synonym* **over-square**) (*the opposite of* **prolate**).
obligate — restricted to a particular host, **substratum**, or means of nutrition, as in the obligate **epiphyll** *Ephemeropsis* (*the opposite of* **facultative**).

Pohlia nutans **nutant** capsules.
1 mm

Ptychomnion aciculare **nutant** shoot apex.
1 mm

Tortula muralis **oblate** basal leaf cells.
100 μm

oblique — having a slanted base, **insertion**, mouth, or tip (•below, right).
obloid — a solid with an **oblong** profile.
oblong — nearly **rectangular** in outline (•below, left and middle).

Syntrichia princeps **oblong** leaf.
0.5 mm

Camptochaete arbuscula **oblong** capsules.
1 mm

Funaria hygrometrica **oblique** operculum.
1 mm

obovate — shaped like an egg but upside down.
obovoid — said of a solid that's egg-shaped in form but is upside down (•below, left).
obscure — dark, indistinct, or hard to see, as with leaf cells obscured by dense surface **papillae** or cell walls obscured by vacuoles or other cytoplasmic inclusions (•below, middle and right).

Tortula truncata **obovoid** capsules.
1 mm

Funaria hygrometrica **obscure** leaf cells.
50 μm

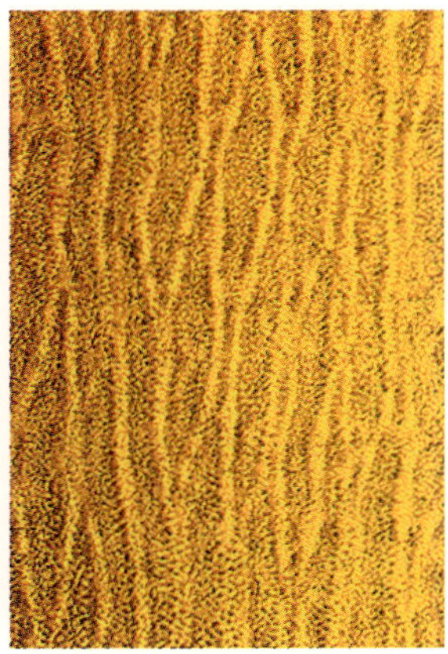

Rhacocarpus purpurascens **obscure** leaf cells.
50 μm

obsolete — barely visible, as with a moss leaf **costa** that's very thin or is almost the same color as the cells around it (•below, left and middle).

Porotrichum ruficaule **obsolete** costa.
100 µm

Metaneckera menziesii **obsolete** costa.
100 µm

Radula silvosa single oil-body **occluding** lumen.
100 µm

obtuse — blunt, with the sides making an angle of more than 90° (•below, left and middle).
occluding — filling up or blocking (•above, right).
ocellus (plural **ocelli**) — in leafy liverworts, a leaf cell that's unusually large, often contains one or more larger-than-usual **oil-bodies**, and lacks **chloroplasts** (•below, right).
ochraceous — brownish yellow, the color of ocher, an oxide of iron widely used as a colorant in paints and ceramic glazes.

Distichophyllum microcarpum **obtuse** leaf apex.
1 mm

Asterella australis **obtuse** shoot apex.
5 mm

Rectolejeunea ocellata **ocelli**.
100 µm

ocrea (or **ochrea**, plural **ocreae** or **ochreae**) — in mosses, a thin **sheath** terminating the **vaginula**.

octofarious — arranged in eight rows or **ranks**.

-oid — a suffix meaning *resembling* or *similar*, for example in **habit** or in a particular trait.

oil-bodies — oil-containing structures in the cells of most leafy and **thallose** liverworts. Their size, shape, color, and number per cell are often characteristic of a given **species** or **genus**, so they're useful for classification and identification. Neither mosses nor hornworts produce such **differentiated** oil-bodies (• below).

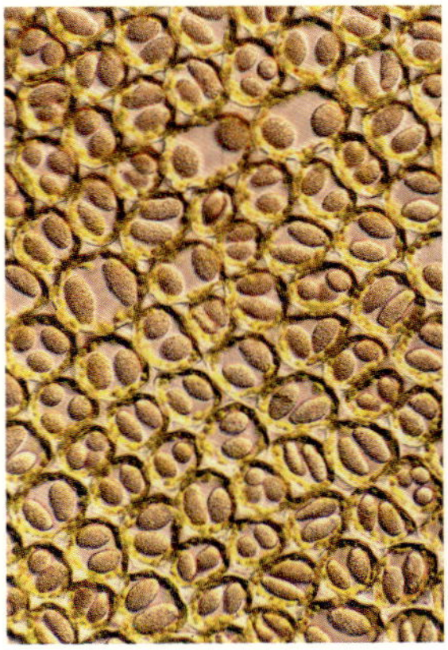

Chiloscyphus sp. **oil-bodies**.
100 µm

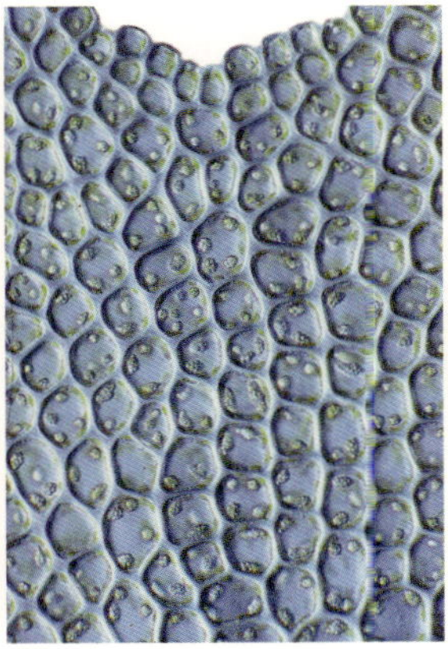

Lepidozia concinna **oil-bodies**.
100 µm

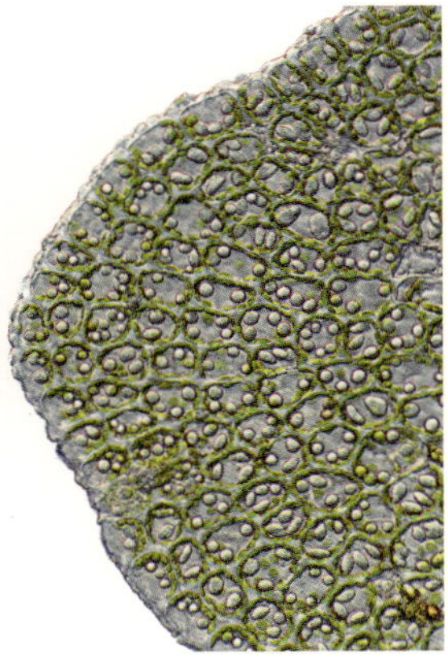

Jungermannia totopapillosa **oil-bodies**.
100 µm

Jackiella curvata **oil-bodies**.
100 µm

Plagiochila sinclairii **oil-bodies**.
100 µm

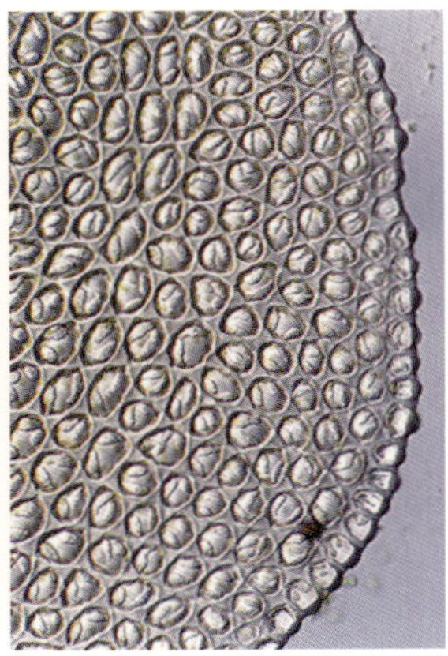

Cheilolejeunea sp. **oil-bodies**.
100 µm

oil-cell — a cell which contains one or more conspicuous oil droplets or **oil-bodies** (•below).
oligo- — a prefix meaning *few*.
oligoporous — having only a few pores.
oligotrophic — poor in mineral nutrients (*compare with* **ombrotrophic**, mineral-poor as a result of getting water from only rainfall, and **eutrophic**, rich in nutrients such as nitrogen and phosphate).
olivaceous — olive-green in color.
ombrotrophic — said of vegetation that gets its water from only rainfall rather than runoff or ground water, and so is mineral-poor (*see also* **bog**).

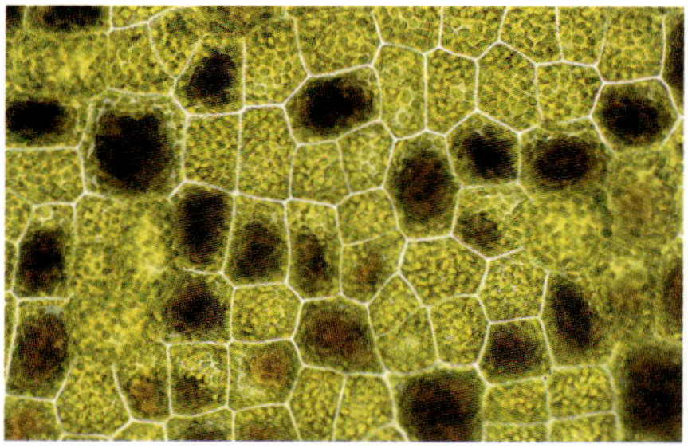

Treubia pygmaea **oil-cells**.
100 µm

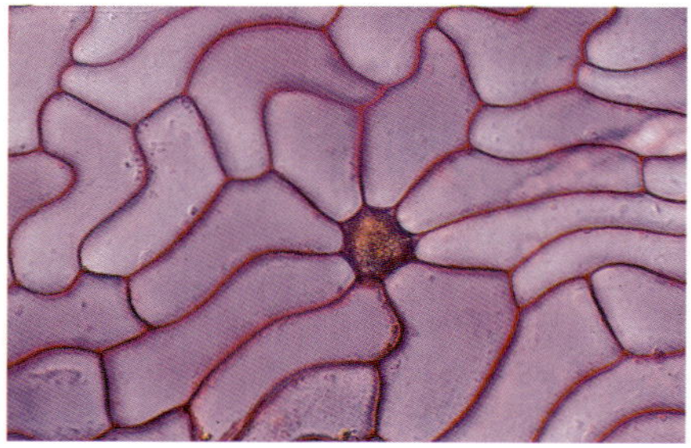

Marchantia foliacea **oil-cell** in a scale.
100 µm

opaque — (1) not allowing any light through, (2) dull or matt on the surface (•below, left).
operculate — possessing an **operculum** (*synonym* **stegocarpous**) (*the opposite of* **astomous**).
operculum (plural **opercula**) — (1) in mosses, the **lid** that blocks the **capsule** mouth (•below, middle and right), (2) in liverworts, the **apical** portion of a **sporangium** which opens during **dehiscence**.

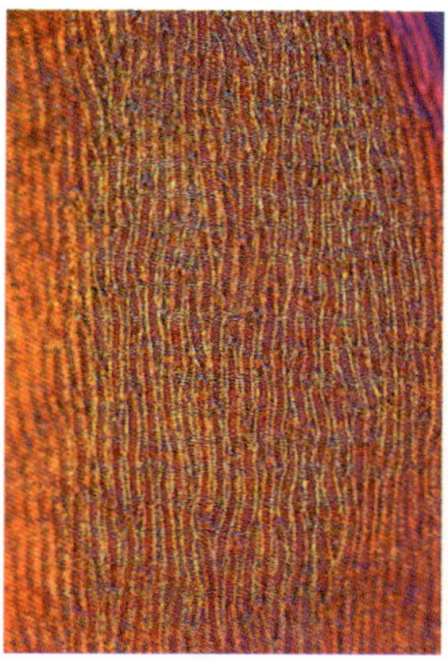

Rhacocarpus purpurascens **opaque** leaf surface.
100 µm

Polytrichadelphus magellanicus **opercula**.
1 mm

Bryum billardierei **opercula** of mature capsules.
5 mm

opposite — said of leaves or branches that are positioned along a stem directly across from each other (•below, left) (*the opposite of* **alternate**).

oral — referring to a moss **capsule**'s mouth (*compare with* **suboral**, referring to the region just below the mouth).

Plagiochilion prolifer **opposite** leaves. 1 mm

Lopholejeunea colensoi **orbicular** underleaf. 100 µm

orbicular — circular in outline or nearly so (•above, right, and below).

Calomnion complanatum **orbicular** dorsal leaf. 100 µm

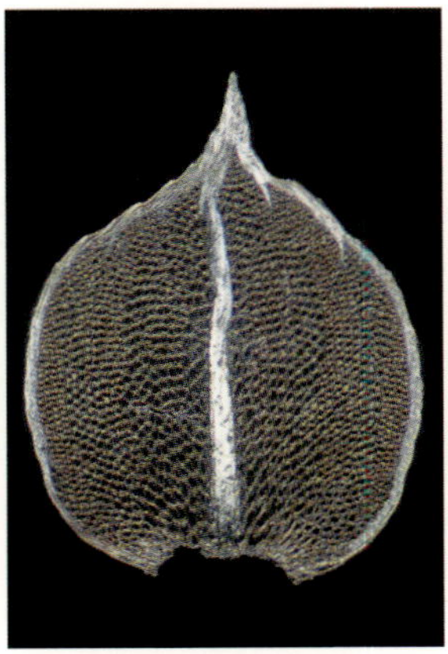

Hypopterygium rotulatum **orbicular** underleaf. 0.5 mm

Rhynchostegium muelleri **orbicular** leaf. 0.5 mm

orgadophilous — preferring a **habitat** of fields or open meadows.

orifice — mouth or opening.

orophilous — preferring a mountainous **habitat**.

orthostichous — said of leaves lined up precisely in a straight row on the stem (*compare with* **distichous**, arranged in two rows, and **tristichous**, arranged in three rows) (*see* **phyllotaxy**).

outer peristome — in mosses, the outer ring of **teeth** of a **double peristome** (•below, left) (*synonym* **exostome**) (*compare with* **inner peristome**) (*see also* **peristome**).
oval — short- or wide-**elliptic** (•below, middle).
ovate — egg-shaped (•below, right).

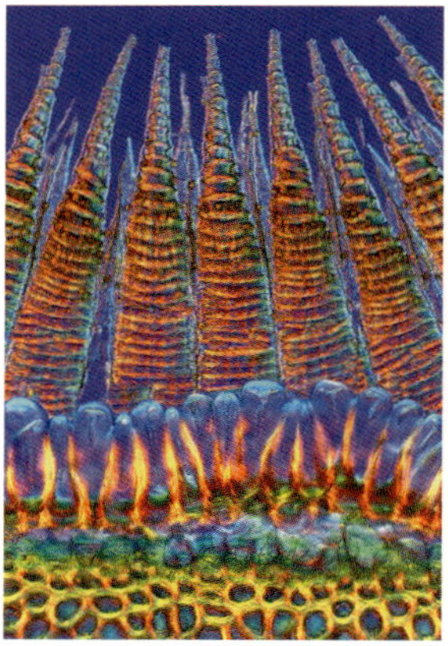

Aulacomnium heterostichum **outer peristome**. 50 µm

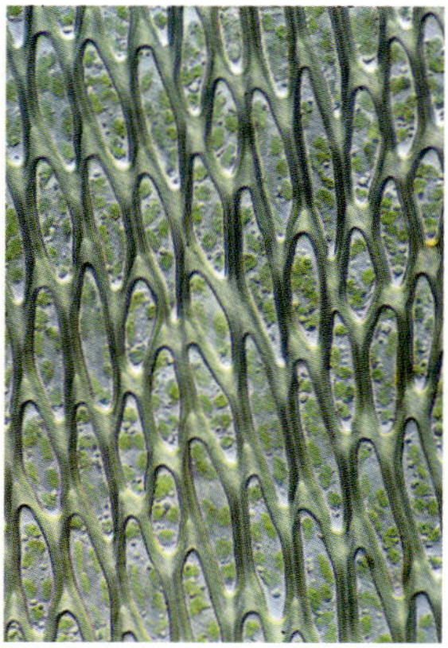

Crosbya straminea **oval** cell lumina. 25 µm

Rhynchostegium tenuifolium **ovate** leaf. 1 mm

over-square — wider than long (•below, left) (*synonym* **oblate**) (*the opposite of* **prolate**).
ovoid — an egg-shaped solid (•below, middle).
ovum (plural **ova**) — an **egg**, the female **gamete** (•below, right).
oxylophilic — preferring an **acidic** or humus-rich **habitat** (*synonym* **acidophilic**).

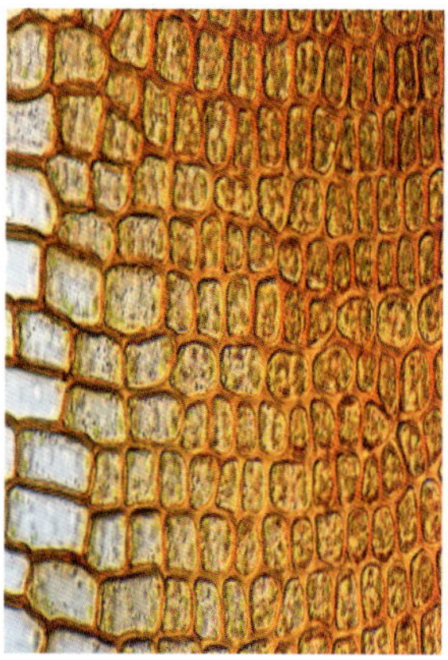

Tortula muralis **over-square** basal leaf cells. 100 µm

Macromitrium longipes **ovoid** capsules. 1 mm

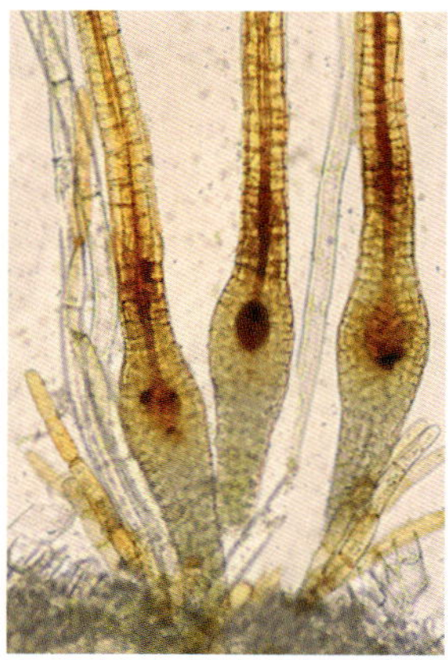

Leptostomum macrocarpon archegonia with **ova**. 50 µm

Camptochaete deflexa **spores**.
20 µm

pachycladous — having thick branches.
pachydermous (or **pachydermatous**) — having a thick skin or wall (*the opposite of* **leptodermous**, thin-skinned or thin-walled).
pachyphyllous — having thick leaves.
palaeotropic — occurring in the Old World tropics.
palea (plural **paleae**) — in liverworts, a chaff-like or membranous **bract**.
paleaceus (or **paleaceous**) — covered with small membranous **scales**.
pallid — pale.
palmate — having radiating finger-like **lobes** (*synonym* **digitate**).
paludicolous — growing in marshes.
palynology — the study of plant **spores** (•left).
panduriform (or **pandurate**) — violin-shaped.
Pangaea — a huge ancient land mass that formed the present-day continents when it broke up.
papilla (plural **papillae**) — (1) a minute, solid protuberance on a cell surface, variously **wart**-like, spinous, **forked**, branched, or shaped like the letter C (•below). A **mammilla** is a bulge in a uniformly thickened cell wall, whereas a **papilla** is a local thickening of the cell wall. (2) a **mucilage**-secreting cell in some liverworts.

Haplohymenium longinerve marginal **papillae**. 100 µm

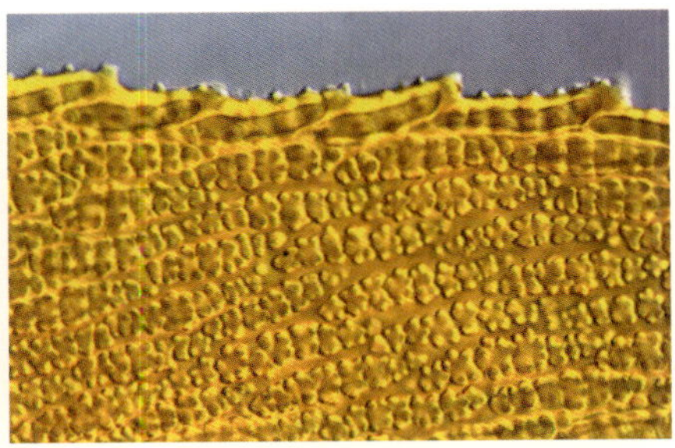

Floribundaria aurea ssp. *nipponica* leaf **papillae**. 50 µm

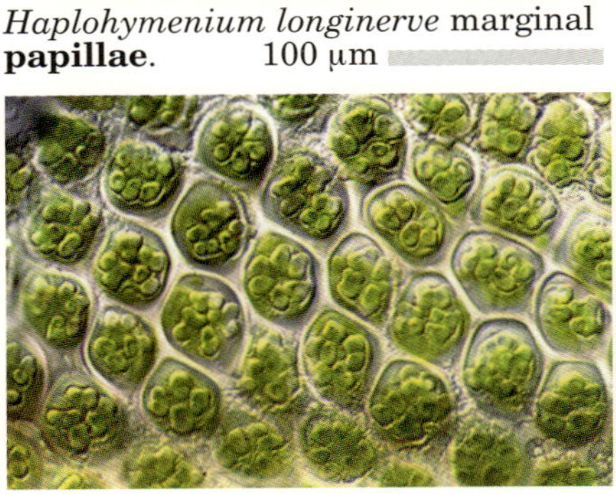

Syntrichia princeps **papillae** on leaf surface. 50 µm

Chiloscyphus leucophyllus **papillae**.
100 µm

papillate (or **papillose** or **papulose**) — roughened or ornamented with one or more **papillae** (*see also* **mammilla**, **prora**, and **scindula**).

papyraceus (or **papyraceous**) — papery (•below, left) (*synonym* **chartaceus**).

paraphyllia (singular **paraphyllium**) — tiny **filaments**, **scales**, or leaf-like structures scattered on the stems of some leafy liverworts and **pleurocarpous** mosses (•below, middle and right).

Pulchrinodus inflatus **papyraceus** leaves.
10 mm

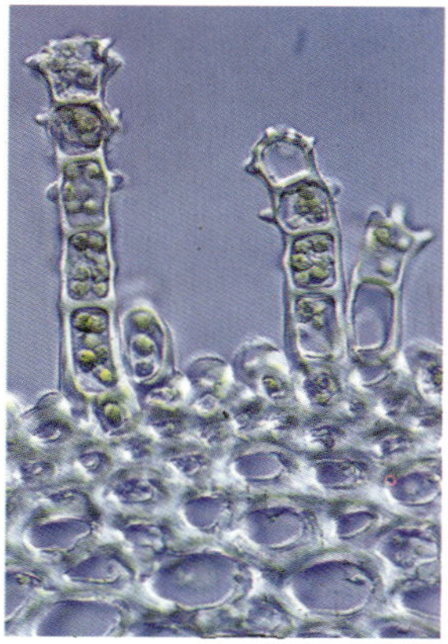

Thuidium laeviusculum **paraphyllia**.
25 µm

Boulaya mittenii **paraphyllium**.
50 µm

paraphyses (singular **paraphysis**) — **sterile**, **septate**, **uniseriate** hairs, sometimes club-shaped or colored, intermixed among the sex organs of most mosses and some liverworts (•below).

Breutelia pendula **paraphyses**.
100 µm

Philonotis scabrifolia **paraphyses**.
50 µm

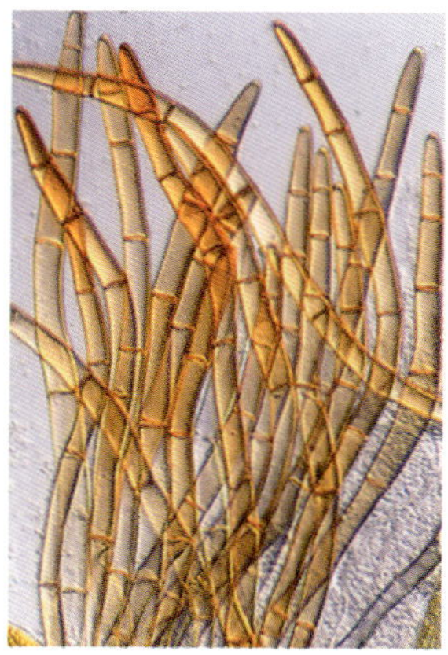

Bryum billardierei **paraphyses**.
50 µm

parasite — any organism or population which improves its reproduction and/or growth at the expense of another. Parasites are no different from **predators** in the *effects* that they have on their hosts, but often they *look* different because most parasites are small and live on or inside their hosts, whereas most predators are large and free-living (*compare with* **mutual**, an organism or population which improves its reproduction and/or growth by living on or near another, which similarly benefits, and **commensal**, an organism or population which improves its reproduction and/or growth by living on or near a host, which isn't affected). Bryophytes suffer from myriad parasites—snails, rotifers, tardigrades, and arthropods browse the leaves of bryophyte gametophores, suck out the contents of leaf and stem cells, or harvest sporangia. However, bryophytes themselves can be parasites when for example leafy liverworts grow luxuriantly enough on host mosses or leaves to reduce photosynthesis.

Polytrichadelphus magellanicus **patent** leaves. 1 mm

parenchyma — a tissue made up of relatively un**differentiated** cells, usually thin-walled and **isodiametric** (*compare with* **prosenchyma**, which has long cells with **tapering** ends that overlap).

parenchymatous — composed of **parenchyma** (*compare with* **prosenchymatous**, composed of **prosenchyma**).

paroicous — having **antheridia** near to the **archegonia** but not mixed in with them, often sited in leaf **axils** just below the **perichaetium** (*compare with* **autoicous** and **synoicous**).

patent — (1) making a 26°–45° angle with the stem (• left), (2) spreading out at 45° or more (*see* **spreading** for a comparison of similar terms).

patulous — making an angle of 46°–90° with the stem (*see* **spreading** to compare similar terms).

peat — deposits of partly decomposed plants, often mostly species of *Sphagnum* (• below).

peat moss — a common name of *Sphagnum* mosses.

Peat polygons in high-arctic Alaska, a product of frost-heaving in permafrost.

Massive **peat** polygons in high-arctic Alaskan permafrost soil.

pectinate — resembling a comb.

pedicellate — stalked.

pediophilous — thriving in flat, open **habitats**.

peg rhizoid — a **rhizoid** with peg-like thickenings on the inner surface of its cell wall (•below, left and middle) (*synonym* **tuberculate** rhizoid).

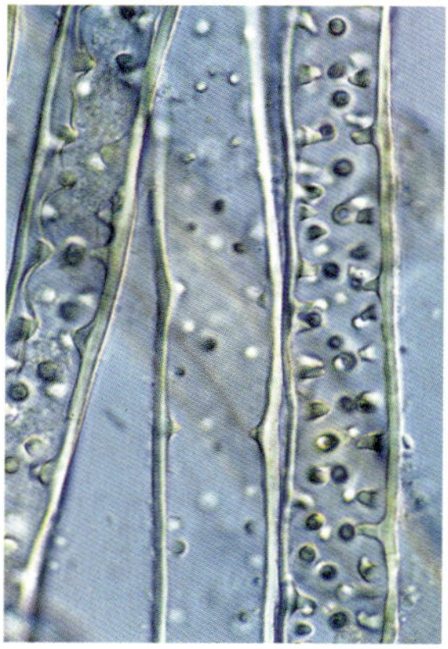

Reboulia hemisphaerica **peg rhizoids** (wm).
50 µm

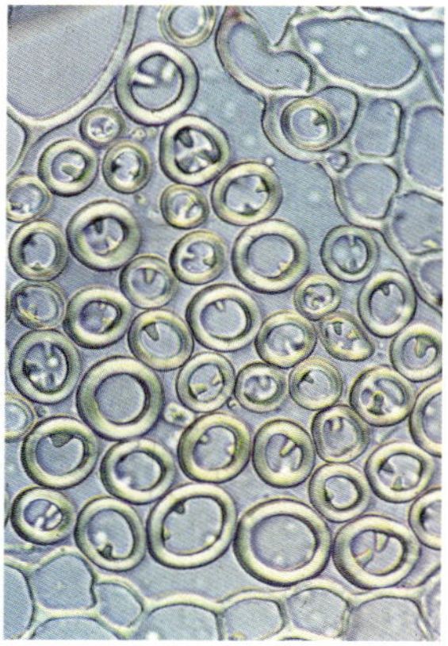

Marchantia macropora **peg rhizoids** (xs).
50 µm

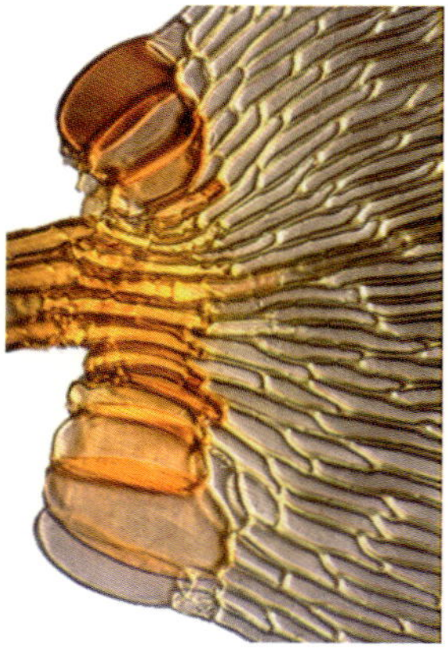

Rhaphidostichum boschii **pellucid** alar cells.
50 µm

pellucid — translucent (•above, right, and below, left).

peltate — shaped like a plate or shield supported by a central stalk (*synonym* **umbilicate**).

pendant — a growth form in which long **gametophores** droop from the **substratum** (•below, middle and right).

pendent — drooping or hanging down.

pendulous — drooping from lack of support.

penicillate — ending in a **tuft** like a brush.

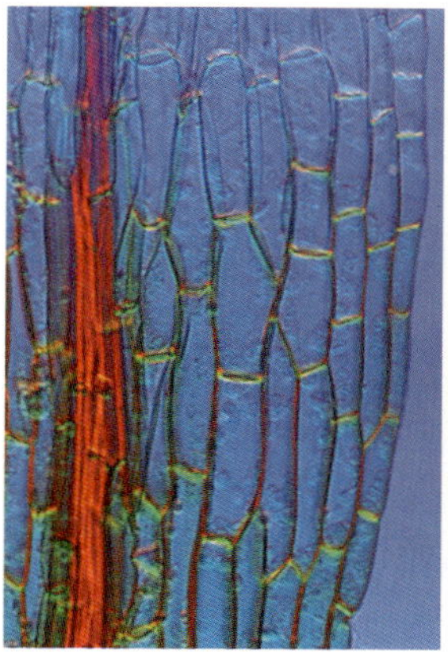

Pseudephemerum nitidum **pellucid** basal leaf cells.
50 µm

Hypnum chrysogaster **pendant** habit.
5 mm

Lejeunea sp. **pendant** habit.
1 mm

pennate — feathery or **pinnate**.

percurrent — extending to the **apex** but not beyond (*compare with* **subpercurrent**, not quite extending to the apex, and **excurrent**, extending beyond the apex) (*see also* **failing**, not reaching the apex).

perennial — a plant that lives for an indefinite number of years, or takes at least three years to mature. Most bryophyte **gametophytes** are perennials, whereas most **sporophytes** are **annuals**, although the sporophyte of the moss *Ptilium crista-castrensis* can take up to three years to mature (*compare with* **annual** and **ephemeral** and **ruderal**, plants that complete their life cycles within a single season and sometimes within only a few months).

perfect peristome — a complete **double peristome** with 16 **teeth** and an **endostome** of 16 **segments** alternating with one or more **cilia** (•below, left) (*synonym* **hypnoid peristome**).

perfoliate — said of a stem that seems to pass right through the blade of a leaf.

perforate — pierced by pores or small holes (*compare with* **fenestrate**, pierced by broad openings resembling windows).

Warnstorfia fluitans **perfect peristome**.
1 mm

Frullania subhampeana mature **perianth**.
1 mm

Acrochila biserialis mature **perianth**.
1 mm

perianth — in jungermannialean liverworts, a **tubular** structure that's formed by the fusion of two or three leaves and protects the developing **sporophyte**. Its size, shape, and surface pleats are useful for identifying species which superficially look similar (•above, middle and right) (*compare with* **pseudoperianth**, in some **thallose** liverworts, a membranous tubular **sheath** that surrounds the sporophyte and is composed of **thallus** tissue).

perichaetial — surrounding the **archegonia** (•below, left and middle).

perichaetial leaf — a modified leaf or **underleaf** surrounding the **archegonia** at the tip of the stem.

perichaetium (plural **perichaetia**) — the female sex organs plus the cluster of modified leaves surrounding them (•below, right) (*compare with* **perigonium**, the male sex organs plus the cluster of modified leaves surrounding them).

Dicnemon calycinum capsules and **perichaetial** bracts. 1 mm

Dicranum plurisetum **perichaetial** bracts. 1 mm

Leptostomum macrocarpon **perichaetium**. 1 mm

periclinal — oriented parallel to the surface (•below, left) (*the opposite of* **anticlinal**).

perigonial bracts — the specialized leaves surrounding the **antheridia** (•below, right).

perigonial leaf — a modified leaf or **underleaf** surrounding one or more **antheridia** in the **axils** of the stem.

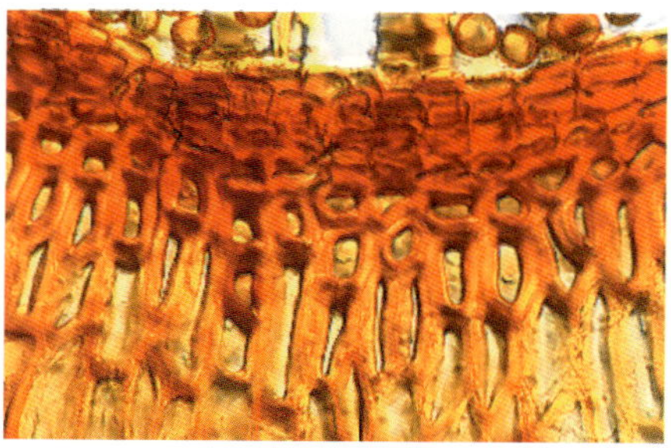

Orthodontium lineare exothecial cells with thick **periclinal** walls. 50 μm

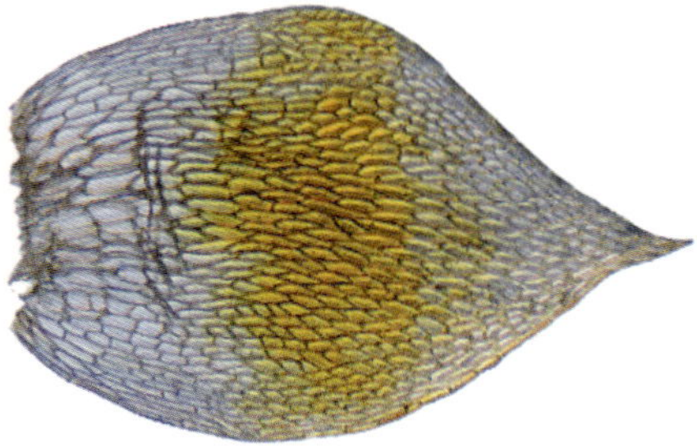

Hypopterygium rotulatum **perigonial bract** (wm). 100 μm

perigonium (plural **perigonia**) — the male sex organs plus the cluster of modified leaves surrounding them (•below except for bottom right) (*compare with* **perichaetium**, the female sex organs plus the surrounding cluster of modified leaves).

Polytrichum juniperinum **perigonia**. 10 mm

Psilopilum crispulum **perigonia**. 1 mm

Catharomnion ciliatum **perigonia**. 1 mm

Breutelia pendula **perigonia**. 1 mm

Lepidolaena clavigera **perigynium**. 1 mm

perigynium (plural **perigynia**) — in some liverworts, a swollen **fleshy** sleeve that's derived at least in part from **thallus** tissue and surrounds the developing **sporophyte** (•above, right) (*synonym* **involucre**).

peristome — a single, double, or rarely multiple circle of **teeth** inside the mouth of a moss **capsule**. A **single peristome** consists of only teeth, whereas in a **double peristome**, the **outer peristome** (**exostome**) consists of **teeth**, and the **inner peristome** (**endostome**) consists of **segments**, sometimes alternating with **cilia**, and often arising from a **basal membrane**.

perpusillate (or **pusillate**) — very small.
persistent — remaining attached, not **deciduous**.
petaloid — resembling a petal.
petrocolous — growing on rocks.
petrophilous — preferring a rock **habitat**.
phaneropore — a **stomate** that's fully exposed on the surface, with its **guard cells** on the same level as adjacent **exothecial** cells (•below, left) (*the opposite of* **cryptopore**, a stomate that's largely hidden, **immersed** in a depression or overarched by adjacent exothecial cells).
photoperiod — (1) the length of daylight, (2) how long a plant is in sunlight during a day.
photoperiodism — the response of living things to the duration and timing of the dark and light periods during a 24-hour period. In some bryophytes, the production of sex organs is triggered by **photoperiod**, but temperature and moisture are always important as well.
photophilous — thriving in well-lighted **habitats** (*synonym* **heliophilous**) (*the opposite of* **sciaphilous**, thriving in shady habitats).
photophyte — a plant adapted to sunny **habitats** (•below, middle and right).

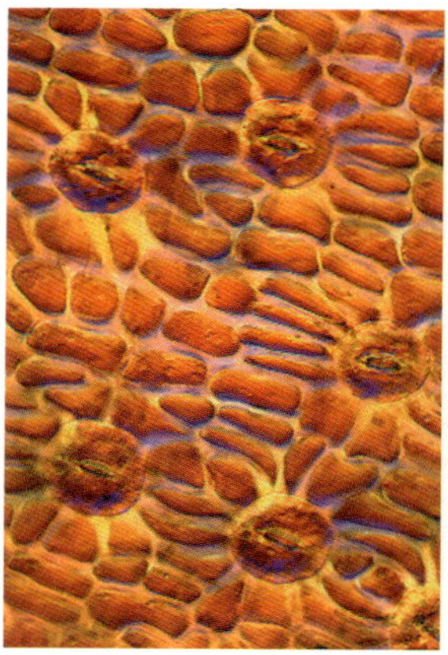

Funaria hygrometrica **phaneropores**.
50 µm

Racomitrium pruinosum **photophyte**.
10 mm

Syntrichia princeps **photophyte**.
10 mm

photosynthate — the energy-rich molecules built from carbon dioxide, mineral nutrients, and the energy of sunlight by the process of photosynthesis and the many metabolic pathways that are "downstream" of it (*synonym* **assimilate**).
photosynthetic — containing the pigment chlorophyll and other metabolic "machinery" for making carbon dioxide energy-rich with the energy of sunlight.

Buxbaumia aphylla mature capsule oriented by **phototropism**.
3 mm

phototropism — the growth response of a plant to directional light. The **capsules** of the **elf cap moss** *Buxbaumia aphylla* (•left) point in the direction of maximum light, a phototropic response that's most likely controlled by **auxin**.

phyllidium (plural **phyllidia**) — (1) any bryophyte leaf, a term that some bryologists favor over *leaf*, (2) a leaf-like **lobe** of *Treubia* and many other metzgerialean liverworts (•below, left).

phyllodioicous — **nannandrous**, with **dwarf male** plants **epiphytic** on the leaves or the **tomentum** of much larger female plants (*see* **nannandrous**).

phyllotaxy — the spiral arrangement of leaves on a stem (*synonym* **dispositio**). A 2/5 phyllotaxy means that five leaves occupy two full spiral turns around the stem.

phylogeny — the evolutionary history of a group of organisms such as a **species**.

phytochrome — a pigment in green plants that triggers light-driven responses. The **germination** of at least some moss and liverwort **spores** is known to involve phytochrome.

piceous — colored a glossy black (•below, right).

Treubia pygmaea **phyllidia**.
1 cm

Fissidens strictus **piceous** shoots.
1 mm

piliferous — having a **hair-point** (•below, left).

pilose — clothed with hairs (•below, right) (*see* **hairy** for a comparison of similar terms).

Racomitrium pruinosum **piliferous** shoots. 1 mm

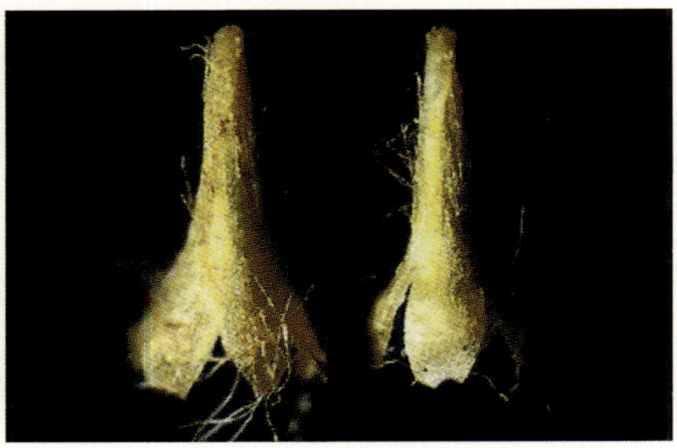

Racopilum robustum **pilose** calyptrae.
1 mm

pincushion moss — (1) the common name given to the moss genus *Leucobryum* because its dense, rounded **cushion** looks like a gigantic pincushion (the tiny stuffed pillow that a seamstress sticks pins into so that she can pick them up quickly with only one hand). The common name is apt, because just as a real pincushion rolls away if it's knocked over, *Leucobryum* cushions scratched loose by foraging birds survive as drifting "vagrant" **moss balls**. (2) The moss genus *Leptostomum* gets the same common name because its large green cushion when sprouting long-stalked **capsules** looks like a giant pincushion with dozens of pins stuck into it.

pinnate — having a "herringbone" pattern of branches on a stem (•below) (*synonym* **pennate**).

Trichocolea sp. **pinnate** shoot.
1 cm

Pseudoscleropodium purum **pinnate** shoot. 10 mm

pioneer mosses — mosses that are among the first colonizers of disturbed or newly created **habitats** (•below, left and middle).

pit — in plant cells, a natural **pore** or perforation in a wall (•below, right).

Funaria hygrometrica **pioneer moss**.
1 mm

Ceratodon purpureus **pioneer moss**.
5 mm

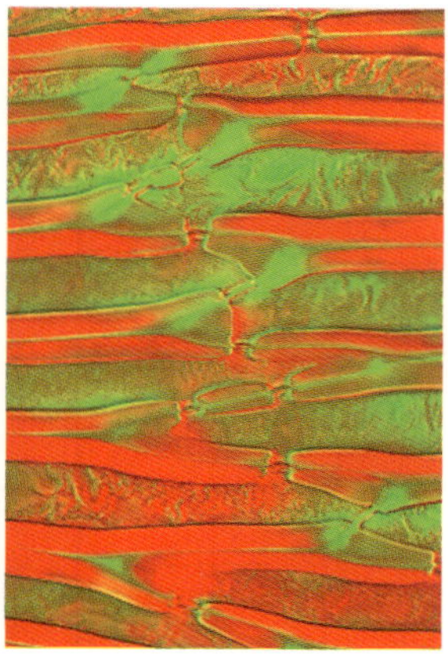

Pulchrinodus inflatus mid-leaf cell **pits**.
10 µm

Lepidozia ulothrix **plane** leaf. 100 μm

pitted — in the walls of plant cell, having **pores** to adjoining cells (•below, right) (*synonym* **porose**).

Spiridens veillardii **pitted** leaf cells. 50 μm

placenta (plural **placentae**) — the junction area between a **gametophyte** and its attached **sporophyte**, usually rich in **transfer cells**.

plane (or **planar**) — (1) flat or level, (2) lying in one plane (•above, left).

planodistichous — lying in one plane and also arranged in two rows on opposite sides of the stem (•below, left).

plasma membrane — the surface membrane of a cell's protoplast (the cell's entire contents inside the cell wall).

plasmodesma (plural **plasmodesmata**) — minute protoplasmic threads that directly connect two adjacent cells' protoplasts (the entire contents of those cells inside their cell walls).

pleated — longitudinally folded (•below, middle and right) (*synonym* **plicate**).

Plagiochila fruticella **planodistichous** shoot. 1 mm

Macromitrium longipes **pleated** calyptrae. 1 mm

Breutelia elongata **pleated** leaves. 1 mm

pleurocarpous — in mosses, producing the female sex organs (**archegonia**) on short side-branches rather than at the tips of the stems (•below) (*the opposite of* **acrocarpous**). As a result, the **sporophytes** are **lateral** rather than **terminal**. Most pleurocarpous mosses are **prostrate**, **matted**, and highly branched, whereas most **acrocarpous** mosses are **erect**, **tufted**, and sparingly branched.

Cladomnion ericoides **pleurocarpous** moss. 2 mm

Cyathophorum bulbosum **pleurocarpous** moss. 1 mm

Cryphaea tenella **pleurocarpous** moss. 1 mm

plica (plural **plicae**) — a lengthwise pleat, fold, or ridge.

plicate — folded lengthwise, forming pleats, folds, or furrows (•below) (*synonym* **pleated**) (*compare with* **sulcate**, strongly plicate, and **striate** or **striolate**, weakly plicate).

plication — a longitudinal pleat, fold, or ridge.

ploidy level — the number of complete sets of **chromosomes** in the cells of an organism. A cell is said to be **haploid** (n) if it contains only a single set, and **diploid** (2n) if it contains two sets. Typically, the **gametophytes** of bryophytes are haploid and the **sporophytes** are diploid (*see also* **polyploid** and **polyploidy**).

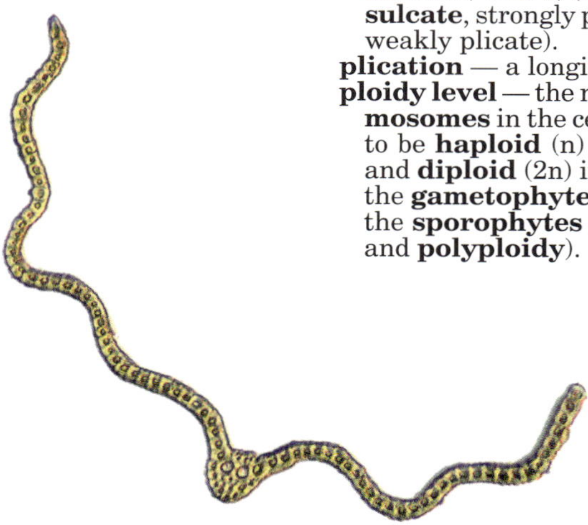

Breutelia pendula **plicate** leaf (xs). 100 µm

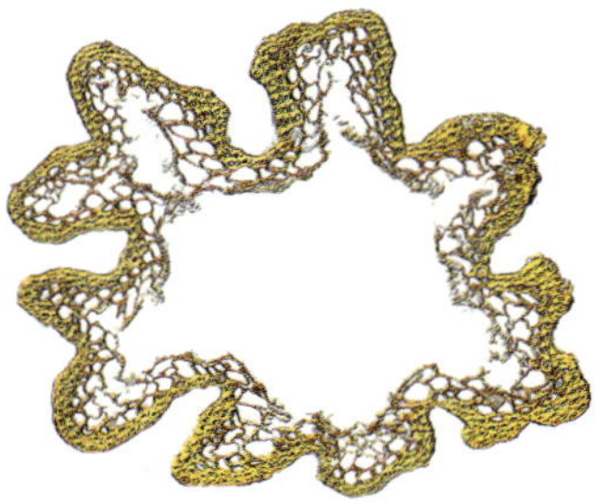

Macromitrium longipes **plicate** calyptra (xs). 100 µm

plumose — feathery (•below, left and middle) (*see* **hairy** for a comparison of similar terms).

Ptilium crista-castrensis **plumose** shoot. 1 mm

Herzogianthus vaginatus **plumose** frond. 1 mm

Dicranum plurisetum **pluri**setose shoot. 1 mm

pluri- — a prefix meaning *several* or *many*, as in plurisetose, producing several **setae** (•above, right), and plurinervose, several-nerved (•below, left).

pluripapillose — having several **papillae** per cell (•below, middle).

pluriseriate — arranged in several rows.

pluristratose — in several layers (•below, right).

polar — in some liverworts and hornworts, having distinct poles like the north and south poles of the earth. In a **triradiate spore**, one pole is the **apex** of the three-angled face, and the other pole is the rounded face opposite it (*the opposite of* **apolar**).

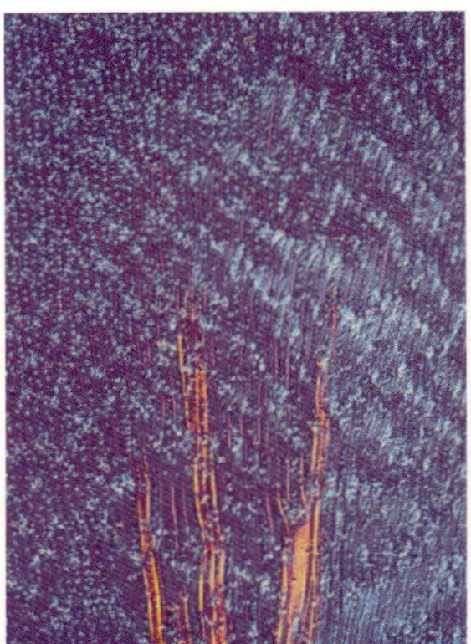

Lepyrodon lagurus **pluri**nervose leaf base. 100 µm

Syntrichia princeps surface of **pluripapillose** leaf. 10 µm

Isolembidium anomalum **pluristratose** leaf base. 100 µm

poly- — a prefix meaning *many* or *numerous*, as in **polymorphic**, having many forms, or polystichous, arranged in many rows, ranks, or series, and polychromic, many-colored.

polyedaphic — growing on many soil types.

polyembryony — in mosses, the production of more than one **sporophyte** from a single **archegonium**, with the sporophytes all sharing a single **calyptra** (*compare with* **polysety**, where each sporophyte has its own calyptra).

polygamous — said of a **species** that produces various types of **inflorescences** on different plants or on the same plant, so that it confusingly seems to be both **dioicous** and **monoicous** (*synonyms* **heteroicous** and **polyoicous**).

polymorphic (or **polymorphous**) — having many forms, highly variable (•below).

polyoicous — said of a **species** that produces various types of **inflorescences** on different plants or on the same plant, so that it confusingly seems to be both **dioicous** and **monoicous** (*synonyms* **heteroicous** and **polygamous**).

Hypnum cupressiforme, a highly **polymorphic** moss species. 1 mm

polyploid — an organism or tissue with more than two complete sets of **chromosomes** in the nuclei of its cells. A triploid has three sets, a tetraploid four, a pentaploid five, a hexaploid six, a heptaploid seven, and an octoploid eight, with **ploidy levels** higher than that usually written 9n, 12n, and so on (*see also* **allopolyploid** and **autopolyploid**). Because the life cycle of most bryophytes is dominated by the **gametophyte** generation (*see* **alternation of generations**) and gametophyte cells have only a single set of chromosomes, deleterious mutations are expressed immediately and therefore are a risk to survival. Polyploidy can protect bryophytes from that risk by masking any damaged genes on one set of chromosomes with normal genes on a second set (*see* **alleles**). With such high survival value, polyploidy is likely to be widespread among bryophytes, and it's been suggested that many liverworts are polyploids, having arisen from ancestors that had a basic chromosome count (called **x**) of 4 or 5. As evidence for that suggestion, most modern liverwort species have a count of 8, 9, or 10, whereas the most primitive living species have only 4 or 5. Mosses too are known to form polyploids—the weedy **cosmopolitan** *Physcomitrium pyriforme* in various parts of its range has counts of 9, 18, 27, 36, and 54.

Dicranum platycaulon **polysety**. 1 mm

polyploidy — the state of having more than two complete sets of **chromosomes** in each cell (*see* **polyploid** and **ploidy level**). Probably a quarter to half of the world's plants have arisen as polyploids. The first *artificial* plant polyploids were produced in mosses. Diploid tissue was induced to form **protonemata**, then **buds** and finally mature **gametophores** (which ordinarily are all **haploid**). Fertilization produced **tetraploid sporophytes**, which then were cultured into tetraploid gametophores.

polysety — in mosses, the production of more than one **sporophyte** at the apex of a single branch, each with its own **calyptra** (•left) (*compare with* **polyembryony**, in which all of the sporophytes share a single calyptra).

pomiform — (1) resembling an apple (•below, left), (2) resembling fruit of any kind.

pore — (1) a **pit** or hole in a cell wall, usually opening to the wall of an adjacent cell, but to the exterior in the **hyaline** cells (**leucocysts**) of species of *Sphagnum* (•below, middle), (2) in some **thallose** liverworts, a hole opening into an internal **airspace** and often ringed with one or more layers of specialized cells (•below, right) (*see* **simple pore** and **compound pore**).

Conostomum pentastichum **pomiform** capsules. 1 mm

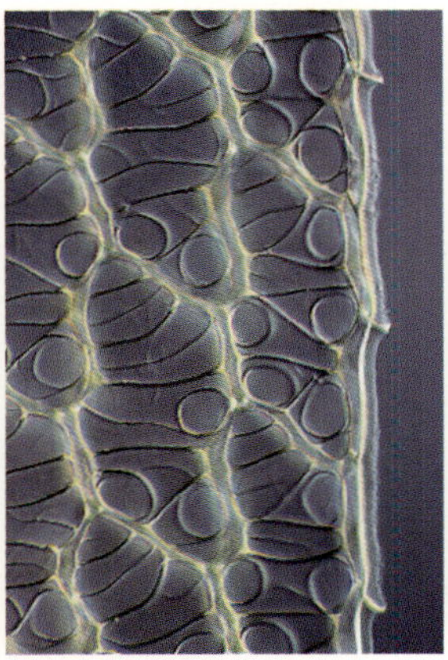

Sphagnum palustre **pores** of hyaline leucocysts. 100 μm

Marchantia foliacea thallus **pores**. 1 mm

porose — having **pits** or **pores**, usually opening to the wall of an adjacent cell (•below) (*see* **pore**).

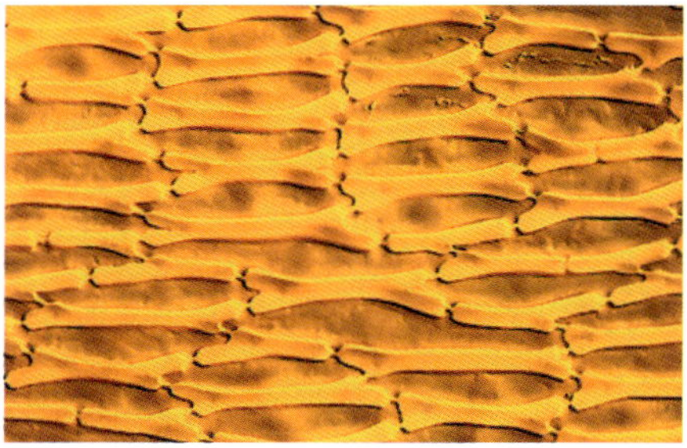

Pseudospiridentopsis horrida **porose** mid-leaf cells. 25 μm

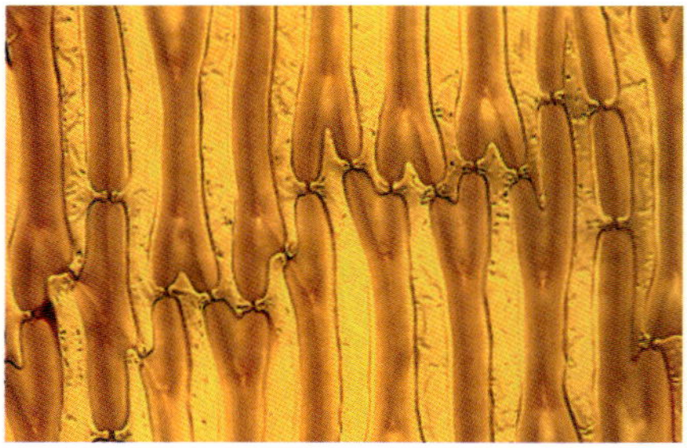

Oedicladium rufescens **porose** mid-leaf cells. 25 μm

postical — (1) facing toward the base (*the opposite of* **antical**, facing toward the **apex**), (2) the side of a **thallus** or stem that faces the **substratum**.

potamocolous — growing in streams (•below, left).

powder gun moss — a common name given to *Diphyscium foliosum* because when its **capsules** are hit by raindrops, they puff out clouds of **spores** like a firing flintlock.

praemorsate — having a ragged **truncate** tip.

precocious germination — **spore germination** preceded by one or more cell divisions inside the spore wall before the spore is shed.

precursor — a structure that precedes or initiates the development of a more complex structure.

predator — any organism or population which improves its growth or reproduction at the expense of another (*see* **parasite**).

predominant — conspicuous or **prominent**.

preperistome (or **prostome**) — irregular thickenings or projections outside the **peristome**.

primary stem — the main stem, often **creeping** and **prostrate** or **rhizome**-like, with **scale**-like or heavily **reduced** leaves (•below, right). Stems branching from it are termed secondary.

Bryum sp. **potamocolous** cushion. 100 mm

Lopidium concinnum **primary stem**. 10 mm

primordial utricle — the collapsed cytoplasmic contents of a cell that have shrunken and pulled away from the cell wall during **desiccation**. A feature of some mosses, the contents often remain collapsed even after re-wetting.

primordium (plural **primordia**) — the embryonic stage of a leaf or other organ, made up of un**differentiated** (unspecialized) cells.

prismaticate — shaped like a prism.

proboscideus (or **proboscideous**) — in the mosses, having a **proboscis**, an extension of the **costa** bearing **gemmae**.

proboscis (plural **proboscides**) — in the mosses, an extension of the **costa** bearing **gemmae**.

procerate — atypically tall.

process (plural **processes**) — in mosses, a **segment** of the **inner peristome** (**endostome**) (•below, left). In most mosses with **double peristomes**, the processes **alternate** with the **teeth** of the **exostome**, but a familiar exception is the genus *Funaria*.

procumbent — spreading over the ground in growth **habit** (•below, middle and right) (*synonym* **prostrate**).

Racopilum aristatum **processes**.
100 μm

Thuidium furfurosum **procumbent** habit.
1 cm

Goebeliella cornigera **procumbent** habit.
1 mm

productivity — the rate of energy capture by an ecosystem or by a particular trophic level within that ecosystem (such as herbivores) (*not the same as* production, which is merely a measure of **biomass**, the total mass of organisms per unit area). Productivity usually is measured in joules or kilocalories per square meter per year, and is said to be *net* rather than *gross* if cellular respiration losses are subtracted.

prolate — longer than wide (•below, left) (*the opposite of* **oblate**, wider than long).

proliferate — to grow or reproduce additional parts or new plants asexually (•below, middle).

Tortula muralis **prolate** basal marginal leaf cells. 100 μm

Leucobryum candidum **proliferating** rhizoids. 1 mm

Hylocomium splendens **proliferous** shoot. 1 mm

proliferous — continuing growth by means of new branch stems (**innovations**) (•above, right).

prominent — conspicuous, such as projecting from a tip or raised above a surface (•below).

Syntrichia princeps **prominent** abaxial costae. 5 mm

Barbula sp. **prominent** costa (xs). 10 μm

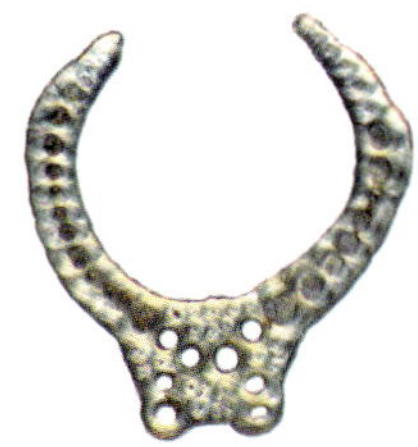

Dicranum plurisetum **prominent** costa (xs). 50 μm

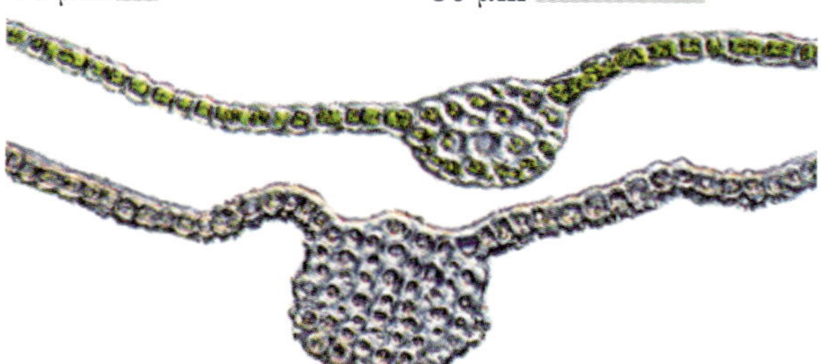

Thamnobryum pandum (top) and *Thuidium laeviusculum* **prominent** costae (xs). 50 μm

propagulum (or **propagule**, plural **propagula** or **propagules**) — a **reduced bud**, branch (**surculus**) or leaf that reproduces a plant by non-sexual means (**vegetatively**) (•below) (*see also* **cladium**, **gemma**, **tuber**, and **brood body**).

Campylopus clavatus stem-apex brood body **propagulum**. 0.5 mm

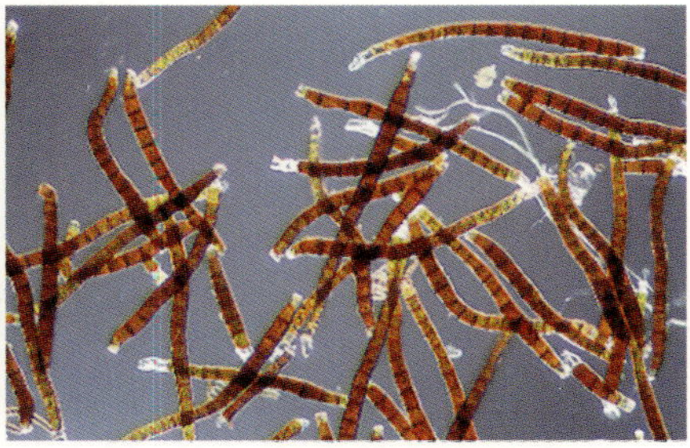

Leptotheca gaudichaudii brood filament **propagula**. 100 μm

prorae (singular **prora**) — bulges in a surface caused by **papillae** or **mammillae** at the tips of cells or by the protruding tips of **prosenchymatous** cells overlapping at their ends (•below, left).

prorate — having a surface roughened by **prorae** (•below, right).

prorulae (singular **prorula**) — minute **papillae** formed by the protruding tips of **prosenchymatous** cells overlapping at their ends (*similar to* **scindulae**).

prorulate — having a minutely **papillose** appearance caused by the protruding tips of **prosenchymatous** cells overlapping at their ends (*similar to* **scindulate**).

prosenchyma — a tissue composed of elongate cells running parallel to the main **axis**, **tapering** and overlapping at their tips rather than joined end-to-end (*compare with* **parenchyma**, a tissue made up of relatively un**differentiated** cells, usually thin-walled and **isodiametric**).

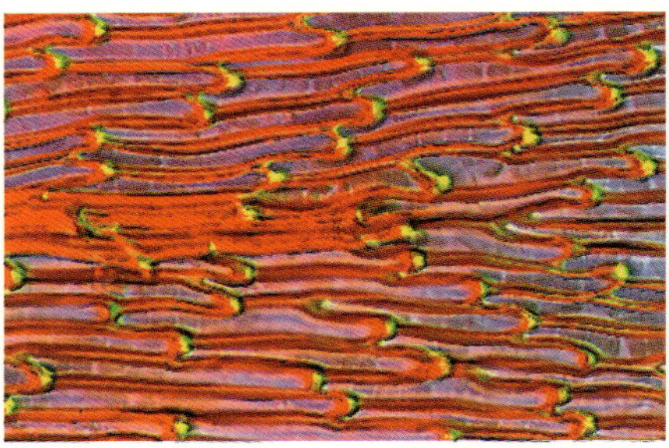

Ectropothecium zollingeri **prorae**. 100 μm

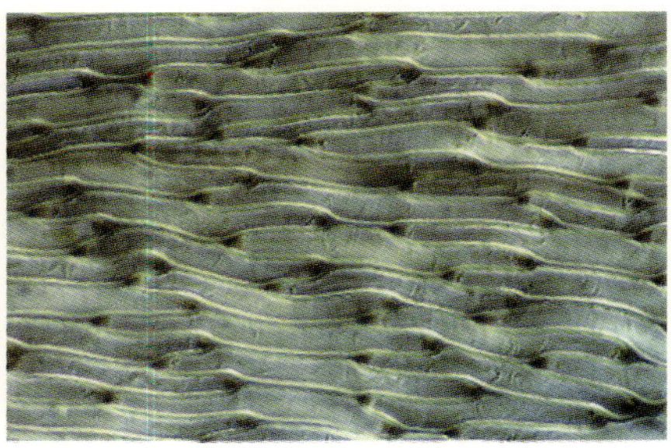

Pleuroziopsis ruthenica **prorate** leaf surface. 25 μm

prosenchymatous — composed of **prosenchyma**, a tissue made up of long, **tapering**, parallel cells overlapping at their tips rather than joined end-to-end (•below, left) (*compare with* **parenchymatous**, composed of relatively un**differentiated** cells, usually thin-walled and **isodiametric**).

prostrate — lying flat on a **substratum** (•below, middle) (*synonym* **procumbent**).

protandrous — said of **monoicous** bryophytes which release their **spermatozoids** before their **archegonia** have matured, thus minimizing the chance of self-fertilization (*the opposite of* **protogynous**).

protean — highly variable (•below, right).

Breutelia arundinifolia **prosenchymatous** leaf. 100 µm

Rhaphidorrhynchium amoenum **prostrate** habit. 1 mm

Ceratodon purpureus **protean** species. 5 mm

protogynous — said of **monoicous** bryophytes which release their **spermatozoids** only after their **archegonia** have matured, minimizing the chance of self-fertilization (*the opposite of* **protandrous**).

protonema (plural **protonemata**) — the juvenile stage of a moss or liverwort, produced by a **germinating spore**, **gemma**, or other **propagule**. Moss protonemata usually are **filamentous**, but they're **thalloid** in species of *Sphagnum* and *Andreaea*. Mature mosses sometimes produce secondary protonemata from leaves, stems, or at wounds. A filamentous moss protonema typically has two stages, the **chloronema** followed by the **caulonema**. A chloronema usually has **hyaline** walls and **transverse** cross-walls, whereas a caulonema usually has pigmented walls and **oblique** cross-walls. Ordinarily, only the caulonema can produce the **buds** which develop into the familiar leafy **gametophores** of an adult moss (•left). Hornworts lack protonemata. Some liverworts produce protonemata, but the walls are thin, colorless, and not oblique, and each produces only a single plant.

Matted moss **protonemata** on clay bank.

protonematal flaps — leaf-like structures on the **protonemata** of species of the moss family Tetraphidaceae. They produce **buds** which develop into **gametophores**.

proximal — located at the end or portion of a structure that's closest to its base or origin (point of attachment) (*the opposite of* **distal**).

pruinose (or **pruinate**) — having a surface **bloom** or pruina.

pseudautoicous (or **pseudomonoicous**) — said of a **dwarf male** moss plant growing **epiphytically** on a female plant (for a comparison with other arrangements, *see* **autoicous**).

pseudo- — a prefix meaning *false* or *resembling but not identical to*, as in pseudoseptum, **pseudopodium**, and pseudoexcurrent (•below).

pseudoannulus (or **pseudannulus**) — in moss **capsules**, a "false" **annulus** that's made up of small uninflated cells and doesn't drop off.

pseudocalyptra — in the mosses, a persistent **epigonium**, or protective envelope, consisting of the **calyptra** and **vaginula**, which remains intact even after the **capsule** has matured, as in species of *Pyramidula* and the Bryobartramiaceae.

pseudodichotomy — "false" branching which is not produced by the division of an **apical** cell, unlike true branching.

pseudoelater — a "false" **elater** produced by most hornworts. Unlike true elaters of most other bryophytes, they're usually multicellular (•below, left).

pseudoparaphyllia (singular **pseudoparaphyllium**) — structures that superficially resemble **paraphyllia** but are found in only **pleurocarpous** mosses and are clustered around the bases of branches or branch buds (•below, right).

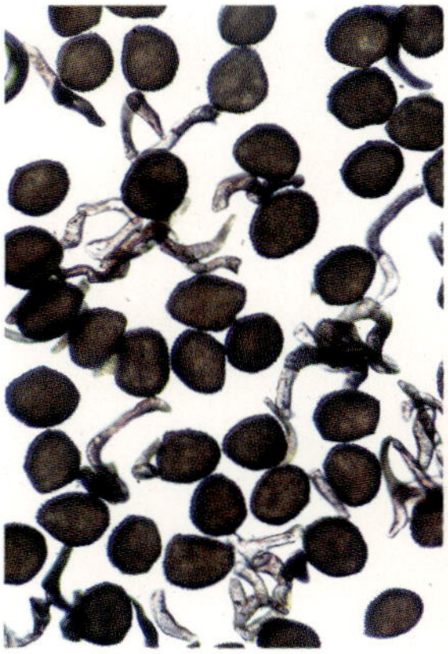

Anthoceros laminiferus **pseudoelaters** and spores. 50 μm

Fissidens bushii **pseudo-**excurrent costa. 100 μm

Taxiphyllum alternans **pseudoparaphyllium**. 50 μm

Asterella australis archegoniophore with **pseudoperianth**. 1 mm

pseudoperianth — in some **thallose** liverworts, a membranous **tubular sheath** that surrounds the **sporophyte**. Unlike a true **perianth**, it's composed of **thallus** tissue and forms after the sporophyte is developed (•left) (*compare with* **perianth**).

pseudopodium — (1) in species of the moss genera *Andreaea* and *Sphagnum*, a tall, leafless, **seta**-like structure that appears to be part of the **sporophyte** because it's topped by a **sporangium**, but in fact is derived entirely from **gametophore** tissue (•below, left)—it pushes the **capsule** beyond the **perichaetial** leaves, thus elevating it just as a true seta does in other mosses, (2) in the moss *Aulacomnium*, an extension of the stem tip bearing **gemmae**.

pseudopore — in the branch leaves of species of *Sphagnum*, a small spot that resembles a **pore** in the **hyaline** cells (**leucocysts**) because it's encircled by **fibril** rings, but in fact is not perforated and therefore doesn't function as a pore.

pseudostoma (plural **pseudostomata**) — in the **capsules** of *Sphagnum* species, **vestigial stomata** made up of two **guard cells** but no **pore**.

pubescent — downy, covered with soft, fine hair (*see* **hairy** for a comparison of similar terms).

pugioniform — dagger-shaped.

pulverulent — powdery.

pulvinate — shaped like a **cushion** (•below, middle and right).

Sphagnum sp. **pseudopodia** with undehisced capsules. 1 mm

Gravel garden featuring **pulvinate** mosses.

Leptostomum inclinans **pulvinate** growth habit (cutaway view). 10 mm

punctate — covered with dots or small impressions resembling pin-pricks.

punctulate — covered with minute dots.

pungent — ending in a sharp and stiff point.

purpureous (or **purpureus**) — purple (• below).
pusillate (or **perpusillate**) — very small.
pustulose — covered with pimple- or blister-like protuberances.
pyriform — pear-shaped.

Ceratodon purpureus mature **purpureous** setae. 1 mm

Reboulia hemisphaerica **purpureous** margin. 5 mm

Marchantia foliacea **purpureous** scales. 1 mm

pyrophilous — fire-loving, said of mosses that are early colonizers of burned-over areas (• below, left).
quadrant — (1) four cells clustered around a central point, (2) the four cells resulting from the first two divisions of a **capsule**'s **differentiation**.
quadrate — appearing square-sided or nearly so (• below, middle and right).

Polytrichum juniperinum **pyrophilous** species. 5 mm

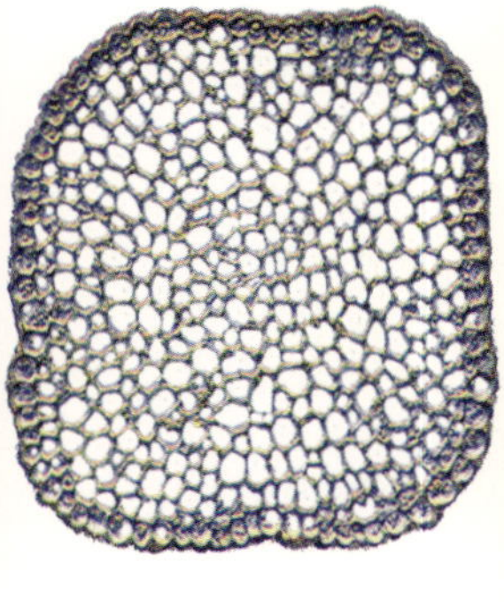

Macromitrium longipes young capsule **quadrate** section. 100 µm

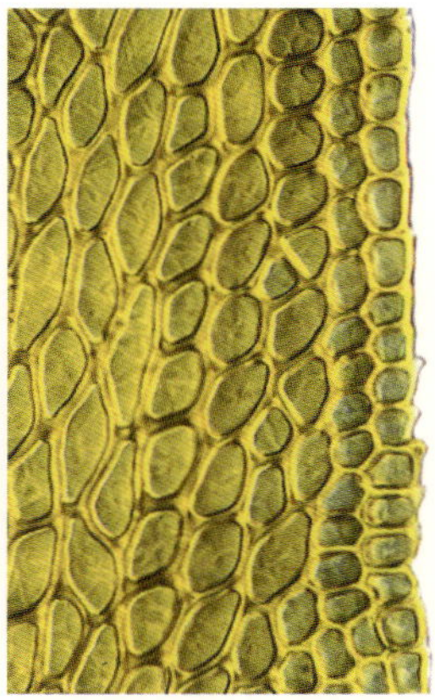

Pinnatella makinoi **quadrate** leaf margin cells. 25 µm

quadrifid — divided into four nearly equal segments (•below).

Lepidozia microphylla **quadrifid** leaf. 100 μm

Kurzia compacta **quadrifid** leaf. 100 μm

quinquefarious — arranged in five **ranks** or rows (•below, left).

radial — arranged **symmetrically** in three or more rows around a central **axis**.

radially symmetric — said of anything which can be cut into two mirror halves along more than one plane (*synonym* **actinomorphic**) (*the opposite of* **zygomorphic** or **bilaterally symmetric**). Radial symmetry is considered less specialized than bilateral symmetry.

radiant — arising from a common center.

radical — arising from the root.

radicelliferous — producing **rhizoids**.

radicle — a slender **filament** that attaches a bryophyte to its **substratum** (•below, right) (*synonym* **rhizoid**). In mosses, radicles are **multicellular** with **oblique** cross-walls, and often branched and colored brown, whereas in liverworts they're unicellular, not often branched, colorless or brown or red. In both groups, they often have **papillae** or **pegs** on their inner or outer surfaces. Hornwort radicles are unicellular, smooth, and colorless.

Conostomum pentastichum **quinquefarious** shoots. 1 mm

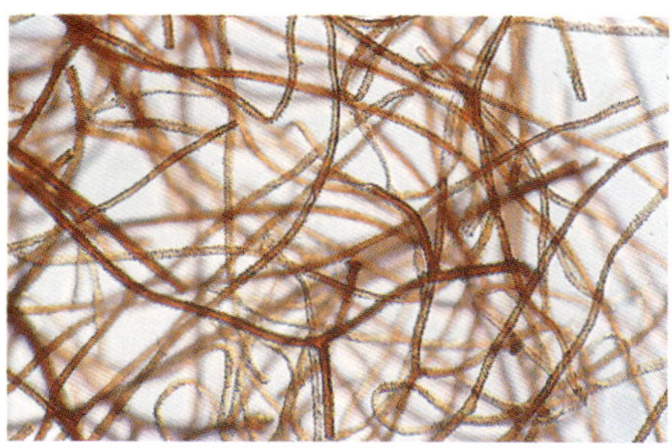

Bartramia pomiformis var. *elongata* massed **radicles**. 100 μm

radiculose — in mosses, covered with highly visible **rhizoids** (•below).
ramentose (or **ramentaceus** or **ramentaceous**) — scaly or fringed with **scales**.

Breutelia pendula **radiculose** shoot.
5 mm

Leptostomum macrocarpon **radiculose** shoots.
1 mm

ramentum — a woolly covering (tomentum) made up of densely matted **radicles** (•below).
rameus (or **rameous**) — part of a branch or belonging to branches.
ramose — much-branched (•left).
ramulose — having many small branches.
ramulus (plural **ramuli**) — the smallest division of a highly branched plant.
rank — a row of similar structures such as leaves on a stem or **teeth** in a **peristome**.

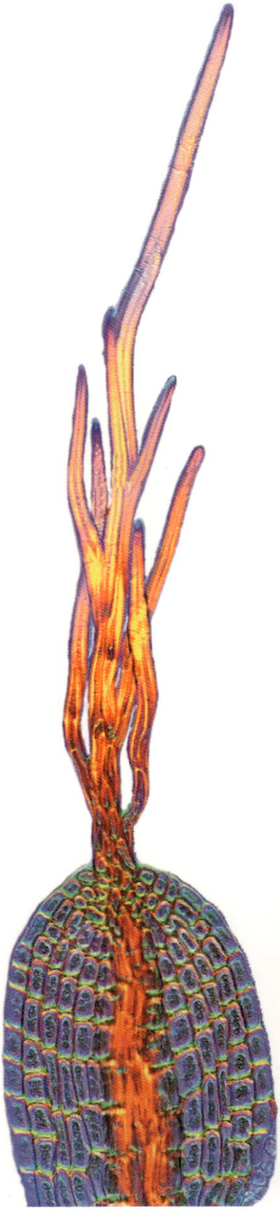

Leptostomum macrocarpon **ramose** hair-point.
100 µm

Campylopus clavatus **ramentum** near leaf bases.
1 mm

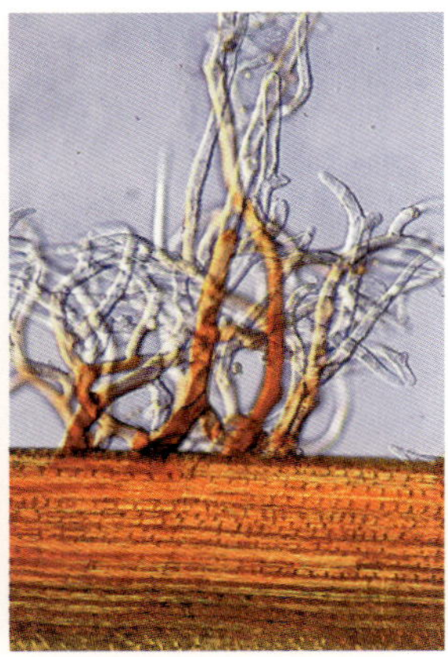

Campylopus clavatus **ramentum** (wm).
100 µm

ranked — arranged in vertical rows.
receptacle — a pad of tissue that bears sex organs (•below, left).
reclined — turned down or bent downwards.
rectangular — 2–3 times longer than wide (•below, middle and right).

Monoclea forsteri male **receptacles**.
5 mm

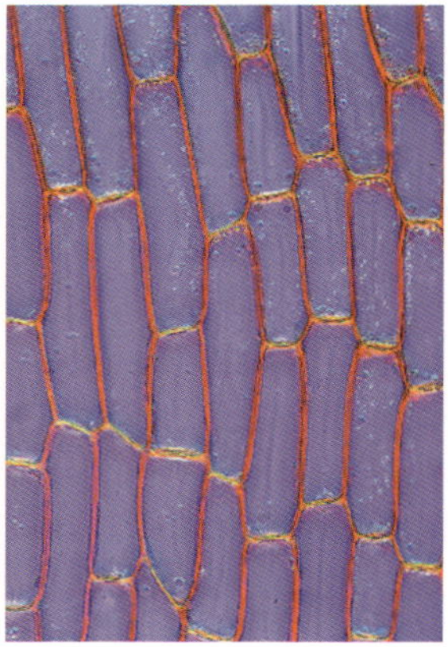

Tortula muralis **rectangular** leaf cells.
100 μm

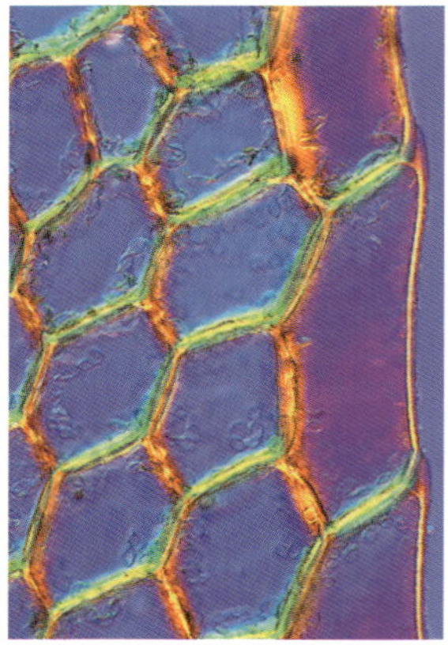

Hookeria acutifolia **rectangular** margin cells.
50 μm

recurved — said of anything that's curved backwards, downwards, or under such as leaf tips and **peristome teeth** (•below, left and middle).
reduced — (1) smaller in size than typical or expected, (2) smaller or structurally simpler than in ancestral stock (•below, right and next page).

Glyphomitrium humillimum **recurved** peristome teeth. 0.5 mm

Tylimanthus saccatus **recurved** leaf margins.
1 mm

Hymenophyton leptopodum **reduced** male branch.
1 mm

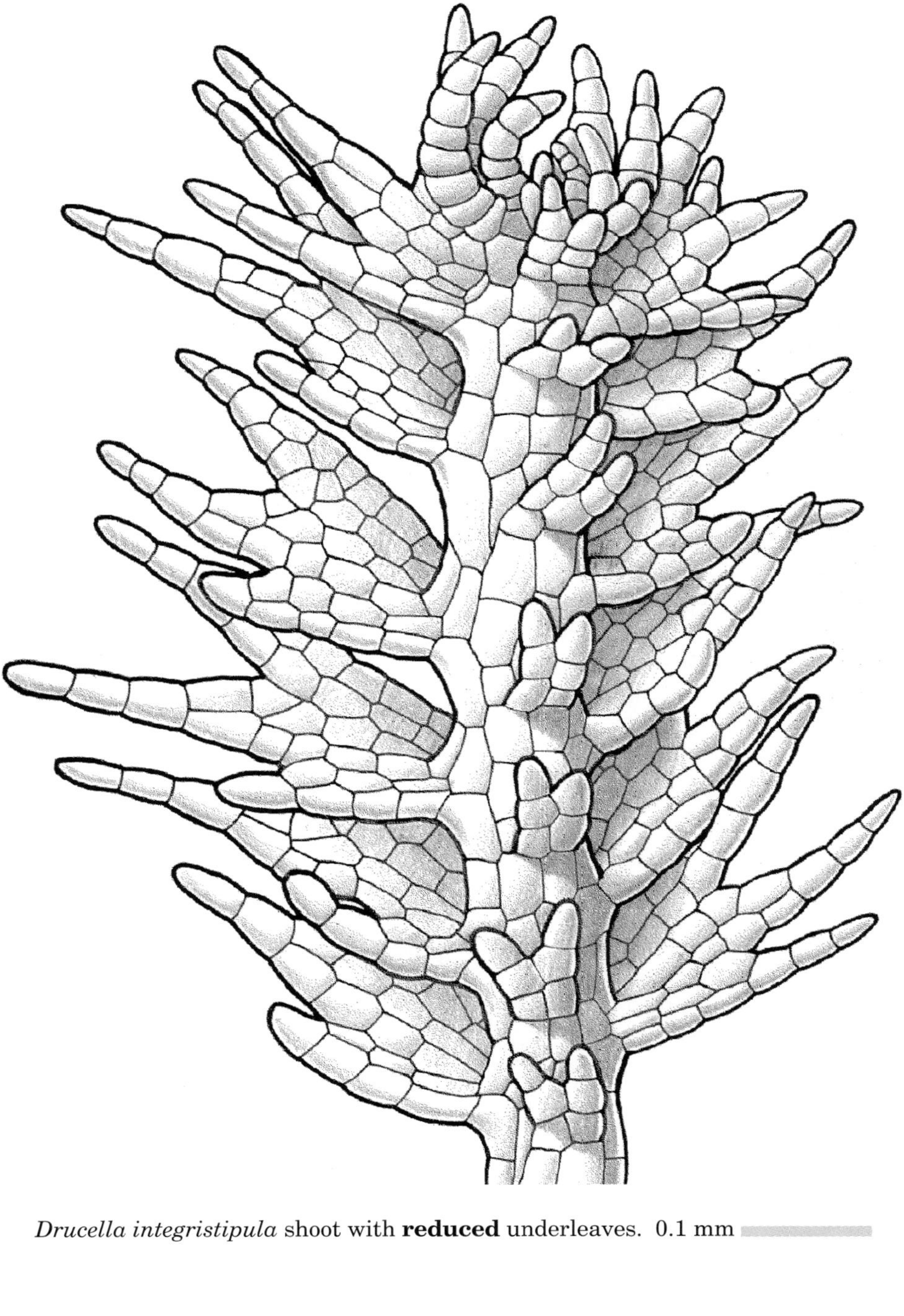

Drucella integristipula shoot with **reduced** underleaves. 0.1 mm

reduction — the loss of an ancestral genetic trait.

reflexed — bent or curved **abruptly** downwards and inwards (•below, left) (*compare with* **inflexed**, abruptly bent or curved upwards or inwards).

regeneration — growth of a new bryophyte plant from a **dedifferentiated** cell that originally was part of a mature structure such as a leaf or stem.

regular — (1) splitting open along a line of built-in weakness such as between **valves** or the rim of an **operculum** (*the opposite of* **irregular**) (*see also* **cleistocarpous** and **stegocarpous**), (2) **radially symmetric** (*the opposite of* **bilateral**), (3) uniform, as in the spacing of **teeth** along a leaf **margin** or leaves along a stem.

Cladomnion ericoides **reflexed** leaf tips.
1 mm

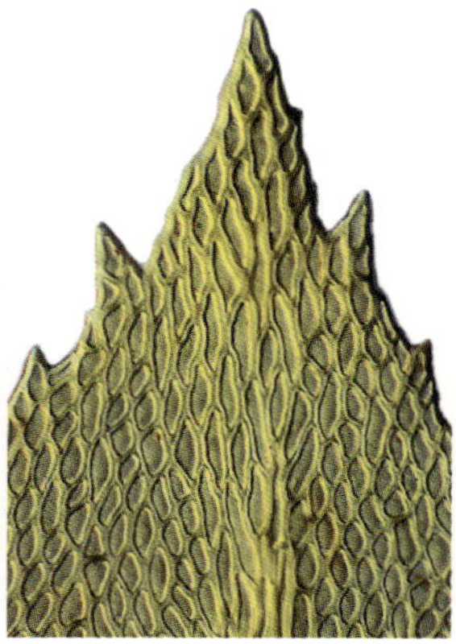

Thamnobryum plicatulum **remote** marginal teeth.
50 µm

Tylimanthus diversifolius **remote** leaves.
1 mm

remote — widely spaced (•above, middle and right) (*synonym* **distant**).

reniform — kidney-shaped (•below).

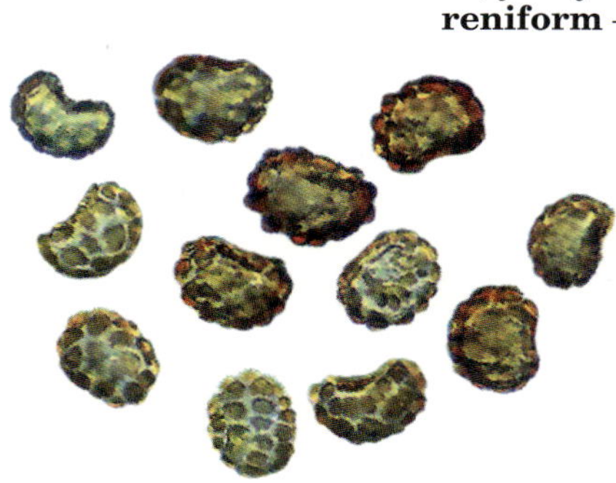

Breutelia pendula **reniform** spores.
50 µm

Phyllothallia nivicola fleshy **reniform** leaves. 1 cm

Macromitrium longipes **repand** lower leaf cells. 25 μm

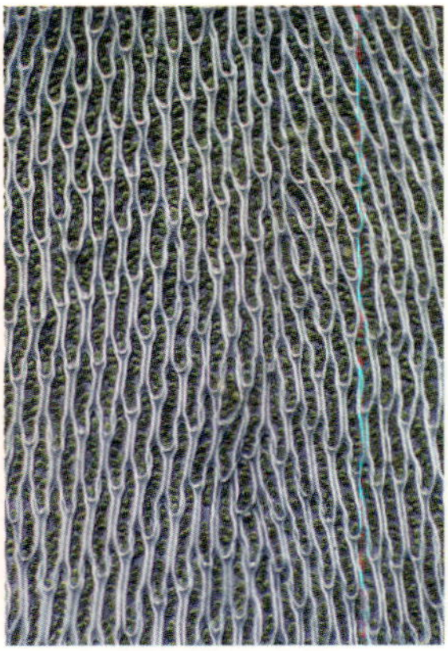

Camptochaete deflexa **repand** leaf cells. 100 μm

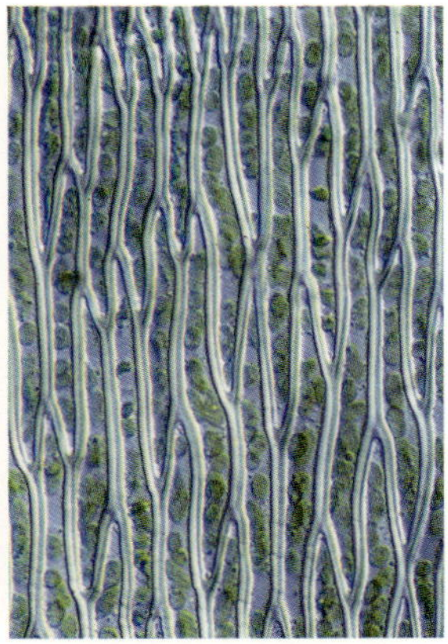

Eriodon cylindritheca **repand** leaf cells. 100 μm

repand — slightly wavy (•above, and below, left) (*compare with* **sinuose**, distinctly wavy).

repent — **creeping** or **prostrate** (•below, right).

resorbed — said of the parts of *Sphagnum* cell walls that break up and erode away.

resorption — the lysing, digestion, or erosion of parts of the cell walls in *Sphagnum* leaves, leaving irregular gaps on either the surfaces of the **hyaline** cells (**leucocysts**) or on the **margins** of the leaves.

resorption furrow — a groove or furrow caused by the erosion of the outer cell walls along the leaf **margins** of some species of *Sphagnum*. Viewed from the surface, the furrow looks like a **border** of fine teeth, and in a cross-section of the leaf, it looks crescent-shaped.

resupinate — said of anything turned upside down by the bending or twisting of its stalk.

reticulate — forming a network.

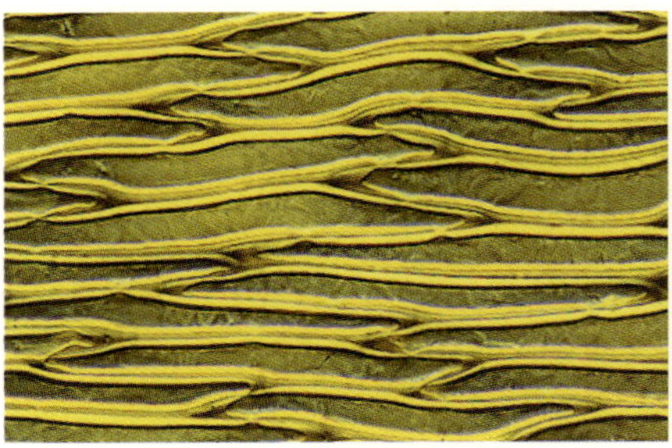

Pleurozium schreberi **repand** leaf cells. 25 μm

Chiloscyphus echinellus **repent** shoots on lignum. 0.5 mm

retort cell — a type of **cortical** cell in some species of *Sphagnum*, shaped like a retort with a projecting neck and ending in a pore that stands out on a **mammilla** near the upper end of the cell.

retrorse — facing or pointing backwards or away from the tip (*the opposite of* **antrorse**, facing forwards or towards the tip) (*compare with* **introrse**, facing forwards or towards the center, and **extrorse**, facing outwards or away from the center).

Dichelodontium nitidum **retuse** leaf apex. 0.5 mm

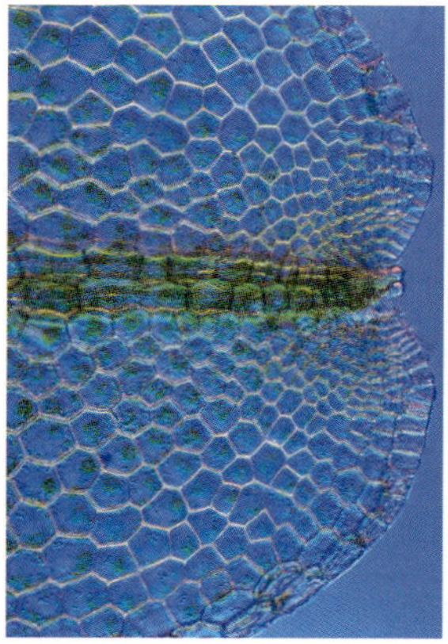

Metzgeria sp. **retuse** frond apex. 100 µm

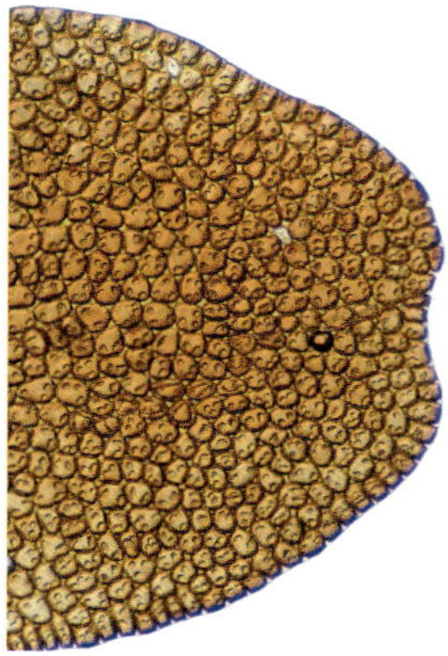

Jungermannia sp. **retuse** leaf apex. 100 µm

retuse — narrowly notched at the **apex** (• above, and below, left) (*synonym* **depressed**) (*compare with* **emarginate**, which denotes a broader notch).

revoluble — rolling away or falling off as a ring (• below, right).

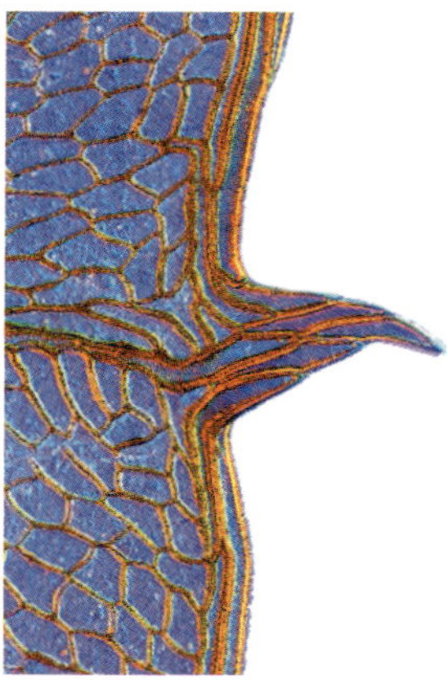

Cinclidium stygium **retuse** leaf apex with apiculus. 100 µm

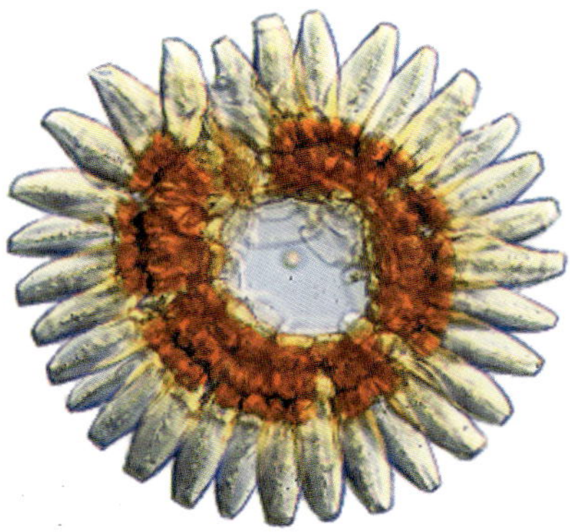

Funaria hygrometrica **revoluble** annulus (detached). 10 µm

revolute — rolled backwards or downwards and under at the edges (•below) (*the opposite of* **involute**, rolled inwards at the edges).

rheophilous — thriving in fast-flowing streams.

rhizautoicous — **monoicous**, with the male **inflorescence** on a short branch attached to the female stem by **rhizoids** (for a comparison with other arrangements, *see* **autoicous**).

Syntrichia princeps **revolute** leaf margin. 50 µm

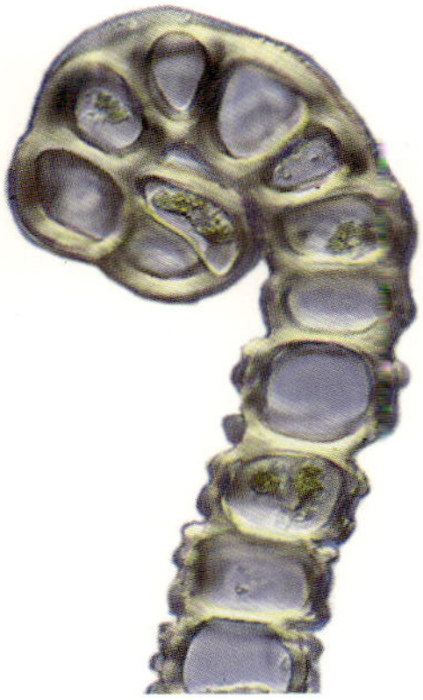

Tortula muralis **revolute** leaf margin (section). 25 µm

Plagiochila circinalis **revolute** leaves. 1 mm

rhizoid — a slender **filament** that attaches a bryophyte to its **substratum** (•below) (*synonym* **radicle**). In mosses, rhizoids are **multicellular** with **oblique** cross-walls, and they're often branched and pigmented (mostly brown), whereas in liverworts they're unicellular, only rarely branched, and colorless or brown or red. In both groups, they often have **papillae** or pegs on their inner or outer surfaces. Hornwort rhizoids are unicellular, smooth, and colorless.

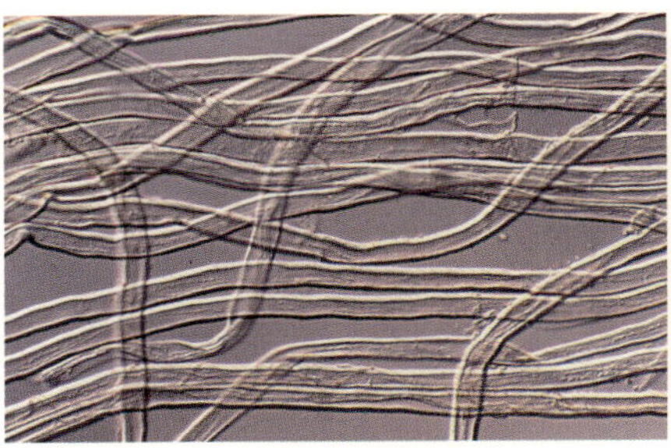

Diplophyllum dioicum **rhizoids**. 100 µm

Fossombronia sp. **rhizoids**. 100 µm

rhizoid furrow (or **rhizoid groove**) — in some marchantialean liverworts, a deep channel running lengthwise in an **archegoniophore** or other structure and containing massed **rhizoids** (•below, left).

rhizoid-initial disk — in leafy liverworts, the pad of tissue at the base of an **underleaf** that gives rise to **rhizoids**.

rhizomatous — having a slender underground stem, horizontal and **creeping**, **analogous** with the **rhizome** of higher plants.

rhizome — a slender, horizontal, underground stem.

rhombic (or **rhomboidal** or **rhomboid**) — shaped like a diamond or rhombus (•below, middle and right).

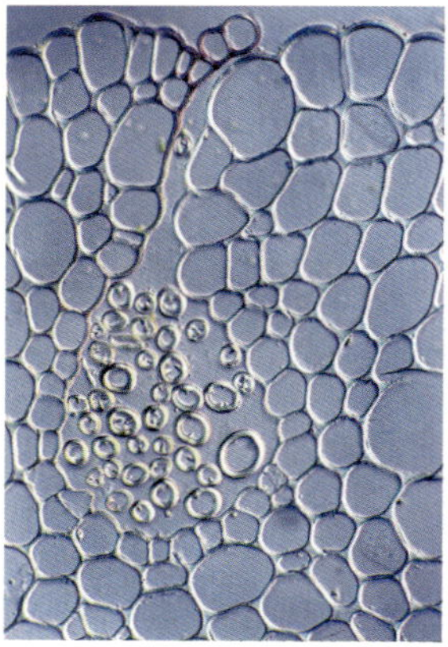

Marchantia macropora **rhizoid furrow** (xs). 100 μm

Homaliadelphus targonianus **rhombic** leaf cells. 100 μm

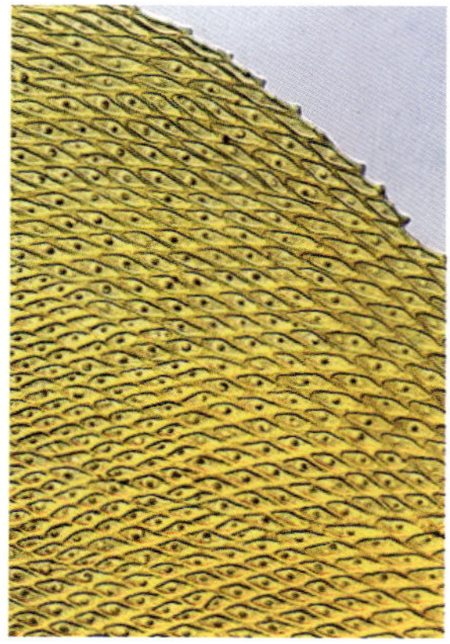

Meteorium buchananii ssp. *helminthocladulum* **rhombic** leaf cells. 100 μm

rimose (or **rimous**) — with irregular surface cracks.

ringed — said of a **pore** surrounded by a **fibril** ring in the leaves of *Sphagnum* species (•below, left).

riparian — of streams and rivers (•below, right).

roridate — appearing to be covered with dew.

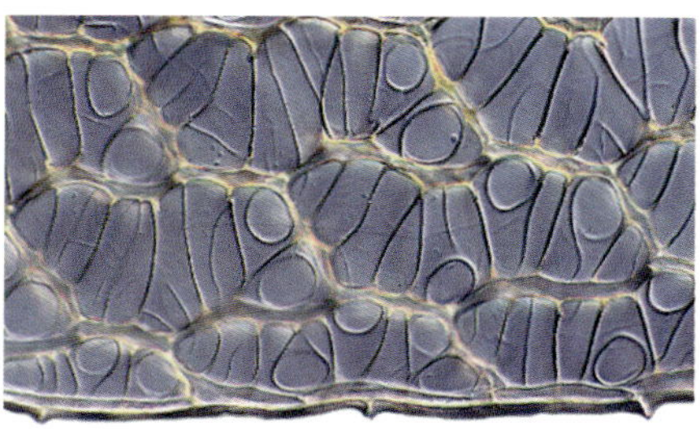

Sphagnum palustre **ringed** pores. 100 μm

Riparian habitat.

Triquetrella papillata **roseus** lower stem.
1 mm

Polytrichum juniperinum **rostellate** operculum.
1 mm

Ptychomnion aciculare **rostrate** opercula.
1 mm

Adelanthus occlusus **rotund** leaf.
0.5 mm

rose moss — a common name given to the moss genus *Rhodobryum* because its **apical** cluster of reddish leaves resembles a rose.

rosette — (1) a cluster of leaves that surrounds a stem (*synonym* **rosula**), (2) a growth form of leaves or **thalli** radiating from a common center.

roseus (or **roseous**) — pale red (•above, left).

rostellate — short-**beaked** (•above, middle).

rostrate — long-**beaked** (•above, right).

rostrum (plural **rostra**) — **beak** (•above, middle and right).

rosula (plural **rosulae**) — a **rosette**.

rosulate — resembling a **rosette**, as in the "male flowers" (**perigonia**) of species of the Polytrichidae mosses (genera such as *Atrichum*, *Dawsonia*, *Polytrichum*, and *Polytrichadelphus*), formed by dense **tufts** of large, crowded, and spreading modified leaves at the tips of the male stems (•below, right).

rotate — shaped like a wheel.

rotund (or **rotundate**) — round (•left and below, left).

Leptodon smithii **rotund** leaf apex.
100 μm

Polytrichadelphus magellanicus **rosulate** perigonium. 1 mm

rubescent — turning red.
rubrous (or **rubous**) — red (•below, left).
ruderal — said of an organism which completes its life cycle within a single season or less (*similar to* **ephemeral**). Most **epiphyllic** mosses, liverworts, and lichens are ruderal, as are **pioneer** species that invade burned or newly disturbed sites.
rudimentary — poorly developed.
rufescent — turning reddish or brownish red (•below, middle).
rufous — reddish or brownish red (•below, right).

Pyrrhobryum mnioides **rubrous** capsule mouth.
1 mm

Polytrichum juniperinum **rufescent** shoots.
5 mm

Ptilidium ciliare **rufous** shoots.
1 cm

rugose — (1) strongly wavy, wrinkled, or undulate **transversely** (*similar to* **undulate**, which means only somewhat wavy, *but in contrast to* **striate**, **plicate**, and **sulcate**, which mean wavy or wrinkled lengthwise rather than crosswise), (2) wrinkled in all dimensions.
rugulose — minutely **rugose** or somewhat rugose (•below, left).
rupestral — growing on rocks (•below, right) (*synonyms* **epilithic**, **petrocolous**, and **saxicolous**).

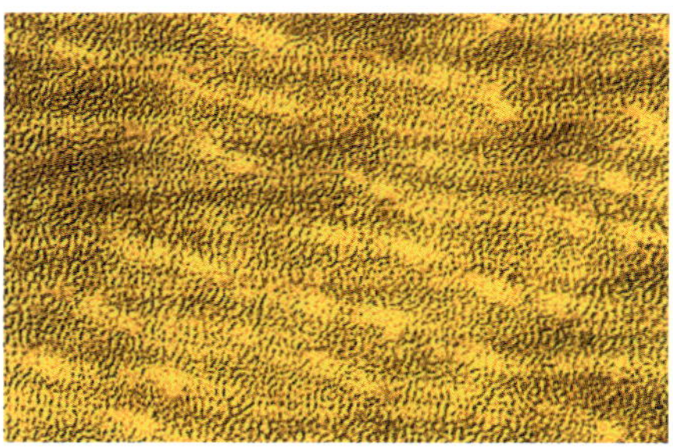

Rhacocarpus purpurascens **rugulose** leaf surface. 25 μm

Macromitrium longipes **rupestral** habit. 5 mm

sabulicolous — growing in sand or gravel.
saccate — forming a sac (•below, left).
sagittate — shaped like an arrowhead.
saprophilous — thriving in humus-rich **habitats**.
saprophyte — a plant that gets its energy directly from non-living organic matter, typically with the aid of endophytic fungi.
savannah (or **savanna**) — grassland vegetation with scattered trees, maintained by occasional fires.
saxicolous — growing on rock (*synonyms* **epilithic**, **petrocolous**, and **rupestral**).
scabrous (or **scabrid** or **scabrose**) — rough (•below, right) (*synonym* **exasperate**).

Austrometzgeria saccata **saccate** leaves. 1 mm

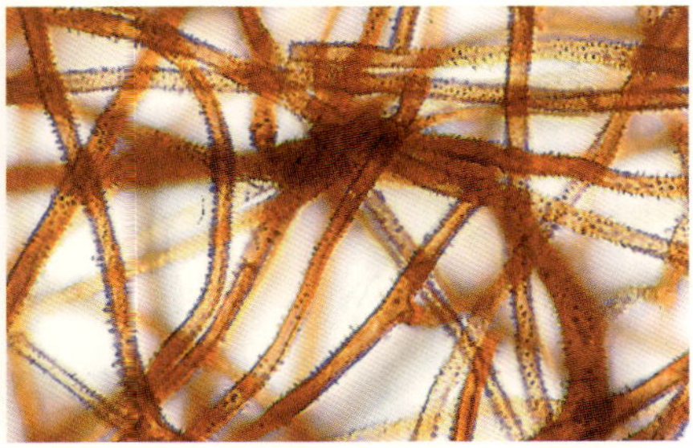

Bartramia pomiformis var. *elongata* **scabrous** rhizoids. 50 μm

scalariform — ladder-like.
scale — a small and membranous leaf-like structure (•below, left and middle).
scalpelliform — resembling the **asymmetric** blade of a penknife or scalpel.
scarious — dry and thin, papery, **scale**-like, or membranous (•below, right).

Marchantia foliacea **scales**. 1 mm

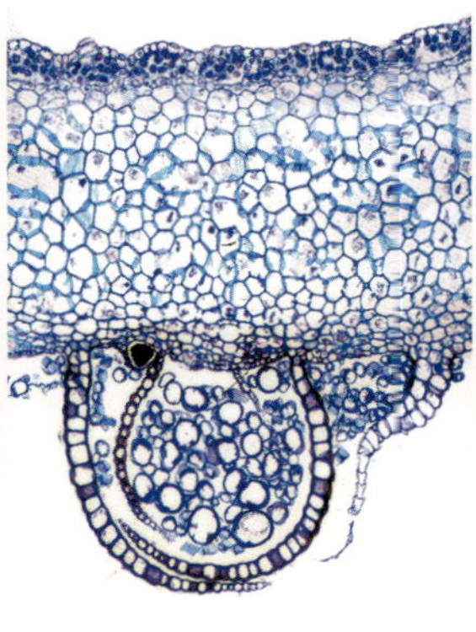

Marchantia sp. **scales** (thallus xs). 0.5 mm

Ptychomnion aciculare **scarious** leaves (dry). 1 mm

sciaphilous (or **sciophilous**) — thriving in shady **habitats** (*the opposite of* **photophilous**).

scindulae (singular **scindula**) — minute **papillae** formed by the protruding tips of cells overlapping at their ends (•below, left) (*similar to* **prorulae**).

scindulate (or **scindulose**) — having a minutely **papillose** appearance caused by the protruding tips of cells overlapping at their ends (*similar to* **prorulate**) (*see* **scindulae**).

scleroderm — a tissue made up of thick-walled cells in the **central cylinder** of the stems and branches of *Sphagnum* mosses and some liverworts.

screw moss — a common name given to the moss genus *Tortula* because its **peristome teeth** are twisted (•below, middle).

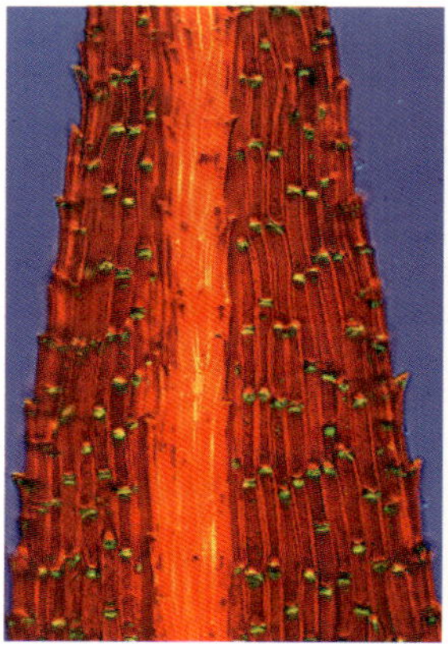

Philonotis tenuis **scindulae** on leaf surface. 100 µm

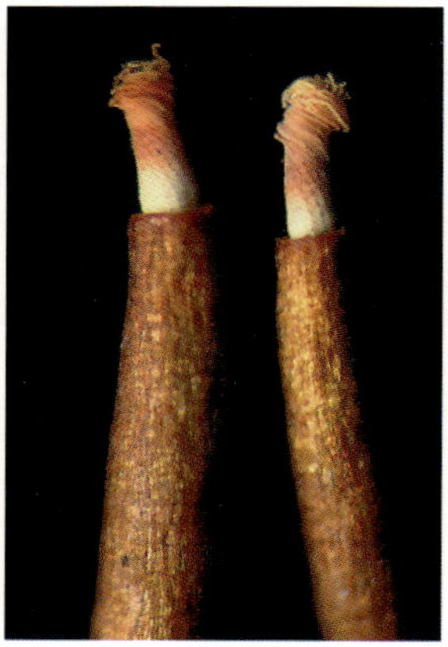

Tortula muralis **screw moss** peristomes. 1 mm

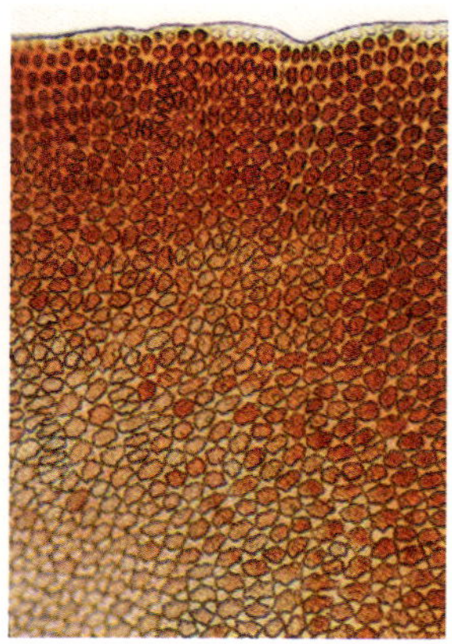

Jamesoniella colorata **secondary pigments**. 100 µm

scrobiculate — pitted with shallow depressions like the surface of a scrotum.

scrotiform — resembling a scrotum.

secondary pigment — a cell wall pigment or a cytoplasm pigment other than chlorophyll (•above, right). As the **seta** of a moss **sporophyte** ages, the walls of the outer cells that make it rigid become red-pigmented. The leaves of bryophytes adapted to full sunlight (for example the moss genera *Andreaea* and *Polytrichum* and the liverwort genera *Ptilidium* and *Frullania*) tend to be pigmented red or brown. The walls of the **spores** of many bryophytes are strongly pigmented, as are the walls of the spore **capsules**. The cell walls of many species of *Sphagnum* are strongly pigmented, and in particular the **antheridial** branches often are red- or brown-pigmented toward their tips.

secund — strongly turned or bent toward one side (•below).

segment — (1) a single tooth-like division of the **inner peristome** (**endostome**) (*synonym* **process**), (2) a discrete globule in an **oil-body**.

Plagiochilion conjugatus **secund** leaves and bracts. 1 mm

Warnstorfia fluitans **secund** leaves. 1 mm

Dicranoloma billardierei **secund** leaves. 1 mm

selliform — shaped like a saddle, drooping on two sides.

semiteres — in cross-section, flat on one side but circular on the other.

sensu — a Latin word meaning *in the sense of*. If it's followed by somebody's name, then it means *in the opinion of* that person, or *as interpreted by* that person. Hence, it's widely used in **Floras** where the treatments of several taxonomists must often be compared.

sensu lato — a Latin phrase meaning *in the broad sense* and either written out in full (in italics) or abbreviated s.l. (*the opposite of* **sensu stricto**—a **genus** taken *sensu lato* includes more **species** than the same genus taken *sensu stricto*).

sensu stricto — a Latin phrase meaning *in the narrow sense* and either written out in full (in italics) or abbreviated s.s. (*the opposite of* **sensu lato**—a **genus** taken *sensu stricto* includes fewer **species** than the same genus taken *sensu lato*).

septate — divided by cross-walls.

septum (plural **septa**) — a wall that separates two end-to-end cells in a **filament**.

seriate — arranged in rows (*compare with* **tiered**, arranged in horizontal or vertical rows).

sericeus (or **sericeous**) — silky, densely covered with fine hairs (*see* **hairy** for a comparison of similar terms).

serrate — regularly toothed like a saw blade, with **teeth** composed of one or more cells and pointing towards the **apex** (•below, middle).

serrulate — minutely regularly toothed with **teeth** composed of only part of a single cell.

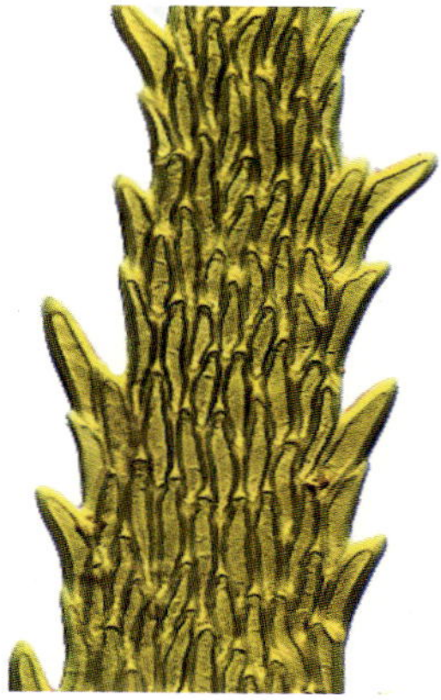

Heterophyllium affine **serrate** leaf margin. 50 µm

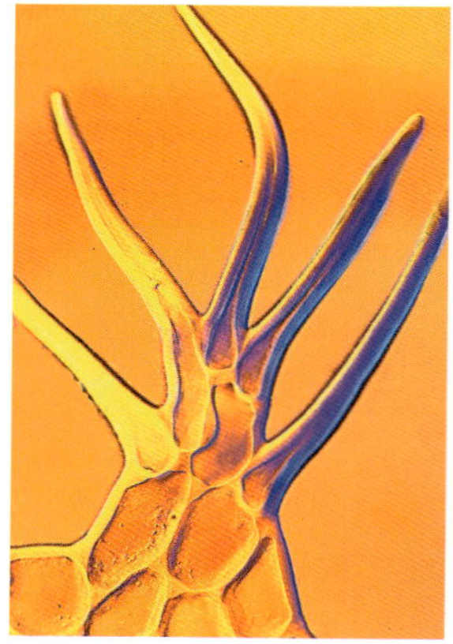

Herzogianthus vaginatus **setaceus** margin. 50 µm

sessile — not stalked, attached directly at the base.

seta (plural **setae**) — the stalk of a moss or liverwort **sporophyte**. Usually the seta of a moss elongates while the sporophyte is developing, whereas that of a liverwort elongates only after the **spores** have matured. The setae of most mosses are wiry, composed largely of rigid thick-walled cells, and photosynthetic when young but later colored with **secondary pigments**, whereas the setae of liverworts are delicate, composed of thin-walled cells that are held rigid by **turgor pressure** and are only rarely **photosynthetic**. The sporophytes of the hornworts always lack setae.

setaceus (or **setaceous** or **setose**) — bristle-like (•left, and above, right).

sheath — a leaf-base that surrounds a stem or **seta**.

sheathing — surrounding or closely clasping a stem, **seta**, or other structure.

shoot — a stem plus leaves and other structures.

shoot-calyptra — in the liverworts, a hollow **fleshy** structure that's derived from both stem and **archegonial** tissue and protects the **embryonic sporophyte**. It's a type of **perigynium** and has the same function as a **perianth** and **bracts** (*synonyms* **stem-calyptra** and **coelocaule**) (*compare with* **calyptra**, an embryo-protecting structure that's derived from only the wall of the archegonium).

shortly rectangular — slightly longer than wide.

Campylopodium euphorocladum **setaceous** leaf apex. 300 µm

shoulder — the **abruptly** narrowed transition between the base and the rest of a leaf.
side-split — said of a **calyptra** that has a single long split (•below, right) (*synonym* **cucullate**) (*compare with* **mitriform**, which is not split up one side and can be **entire** or **lobate** at its base).
sigmoid — S-shaped (•left and below, left).

Funaria hygrometrica **sigmoid** peristome teeth. 50 µm

Zygodon menziesii **side-split** calyptra. 1 mm

siliceous — silica-rich, sandy.
silicicole — a plant growing on sandstone or some other **acidic siliceous** rock.
silicicolous — growing on sandstone or some other **siliceous** rock.
simple — not forked or branched (•below).

Funaria hygrometrica **sigmoid** peristome tooth. 50 µm

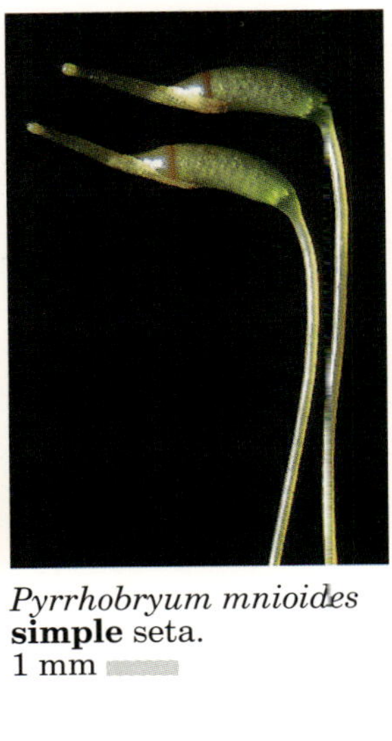

Pyrrhobryum mnioides **simple** seta. 1 mm

Syntrichia princeps **simple** costa extending as awn. 1 mm

simple pore — in the **thallus** of marchantialean liverworts, a type of **pore** that's surrounded by only **superficial** cells (*compare with* **compound pore**, which is composed of an internal barrel-shaped or **cylindric** structure plus several rings of **superficial** cells).

single peristome — in mosses, a **peristome** that's composed of only one ring of **teeth** (•below) (*the opposite of* **double peristome**) (*synonym* **haplolepidous peristome**).

sinistrorse (or **sinistrorsal**) — twisting upwards in a clockwise spiral (from right to left, as in a non-standard left-hand screw thread) (•left) (*the opposite of* **dextrorse**, twisting upwards in a counter-clockwise spiral from left to right, as in a standard right-hand screw thread).

sinuolate — having a minutely wavy or uneven wall or **margin**.

sinuose (or **sinuate**) — having a wavy or uneven wall or **margin** (•bottom of page and next page).

Tortula muralis **sinistrorse** peristome teeth. 100 µm

Schistidium apocarpum **single peristome**. 1 mm

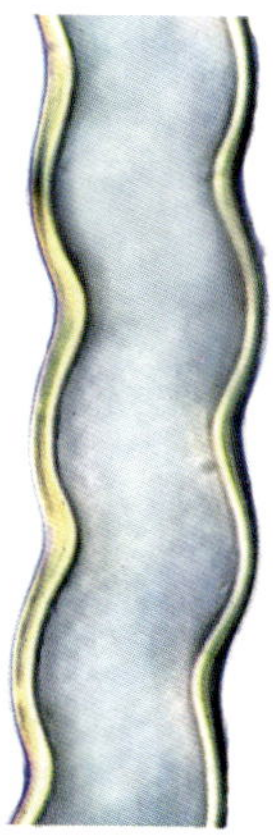

Reboulia hemisphaerica **sinuose** rhizoid wall. 100 µm

Schistidium apocarpum **sinuose** leaf cell walls. 25 µm

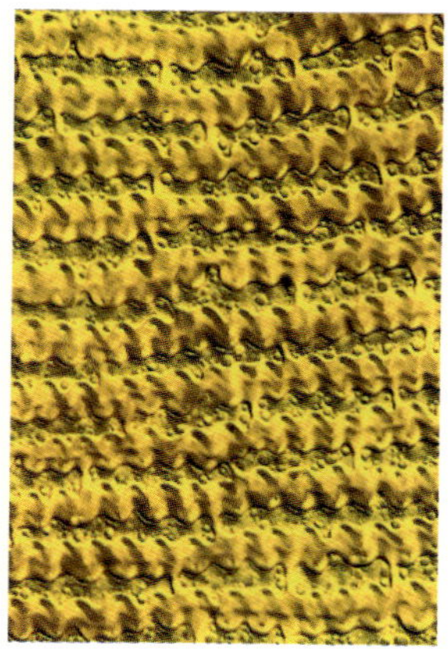

Racomitrium crispulum **sinuose** upper leaf cells. 25 µm

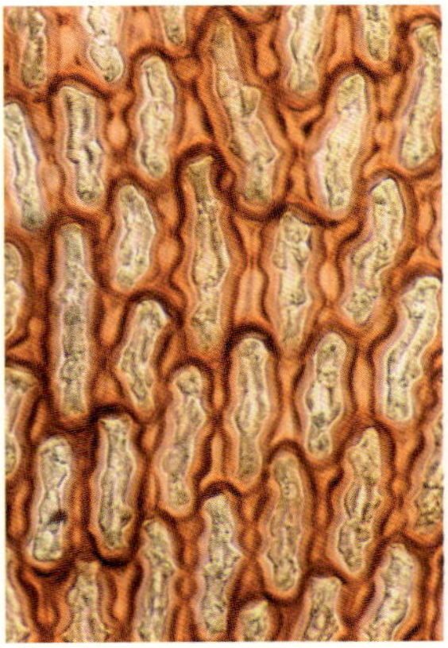

Frullania atrata **sinuose** leaf cell walls.
50 μm

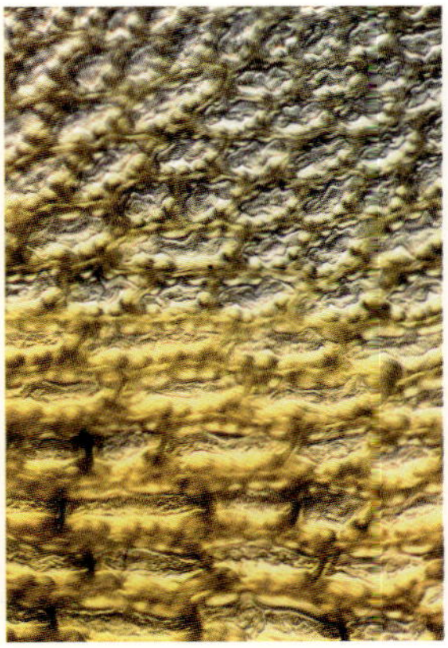

Lepicolea attenuata massive **sinuose** trigones.
50 μm

Verdoornia succulenta **sinuose** thallus margins.
0.5 mm

sinus — the gap between two adjacent **lobes** of a leaf or other structure, typically either U-shaped (horseshoe-shaped, *see* **hyoid** or **ypsiloid**) or V-shaped (*see* **digammoid**) (•below, left).

skiophilous — shade-loving (*an alternative spelling of* **sciophilous** or **sciaphilous**).

slime — a clear, gel-like mucilaginous secretion of some liverworts that's thought to function in absorbing and storing water (•below, middle) (*synonym* **mucilage**).

slime papilla (plural **slime papillae**) — in some liverworts, a cell that secretes mucilage (•below, right) (*synonym* **mucilage papilla**).

Diplophyllum sp. **sinus** (hyoid—U-shaped).
100 μm

Treubia pygmaea **slime** on thallus underside.
5 mm

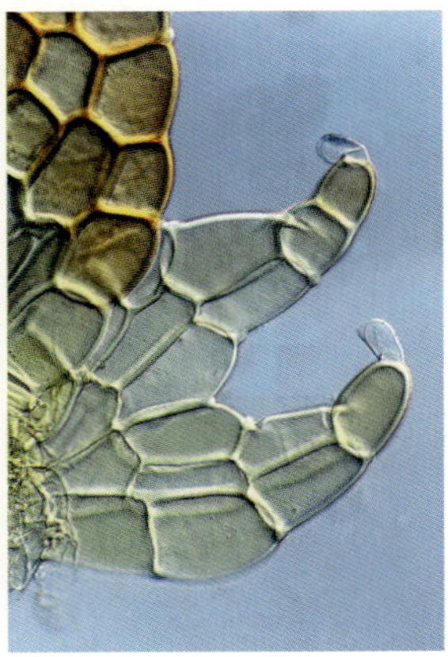

Paracromastigum fiordlandiae **slime papillae**.
50 μm

slime pore — a **stomate**-like **pore** in the **thallus** of some hornworts.

smooth — lacking any surface irregularities such as **papillae**, wrinkles, or ribs (•below, middle and right).

s.n. (abbreviation for **sine numero**) — Latin phrase meaning *lacking a number*, and referring to a herbarium specimen which was never assigned a number by its collector.

somatic cells — the non-reproductive cells of the body of an organism (*compare with* **germ cells**, the reproductive cells of an organism).

sordid — dirty or soiled-looking, colored a muddy or impure hue (*synonym* **squalid**).

spathulate (or **spatulate**) — in the shape of a spatula, narrow below and gradually broadening above (*compare with* **lingulate**, which is less **abruptly** narrowed toward the base).

spear — in mosses, a very young **sporophyte** before the **capsule** has begun to **differentiate** (•below, left).

Spears (from left): *Racopilum convolutaceum, Funaria hygrometrica,* and *Ptychomnion aciculare* (calyptrae removed, various magnifications).

Pseudoscleropodium purum **smooth** leaf cells. 50 μm

Rhacocarpus purpurascens **smooth** leaf border. 50 μm

specialized — (1) said of cells, tissues, or organs that are modified for a particular function, (2) said of organisms or **species** that are adapted to a particular environmental **habitat** or ecosystem **niche**, (3) **differentiated**.

species (plural **species**) — (1) a group of closely related organisms usually interbreeding and sharing many genetic traits, (2) the basic unit of classification of living things, (3) the second word in a binomial (a two-word scientific name), such as *cristatum* of *Sphagnum cristatum*.

specific epithet (or **species name**) — the second word in a Latin binomial (a two-word scientific name), such as *cristatum* of *Sphagnum cristatum*.
spermatid — an immature **spermatozoid**.
spermatozoid — a **biflagellate sperm** (male **gamete**) of a bryophyte (*synonym* **antherozoid**). The sperms of bryophytes are motile and swim to the **egg** through the film of water left by rain or dew.
spiculose (or **spiculate**) — covered with tiny sharp **teeth** or **papillae**.
spindle-shaped — long and narrow and **tapered** at both ends (*synonym* **fusiform**).
spinose — having sharp **teeth** or spines (•below).

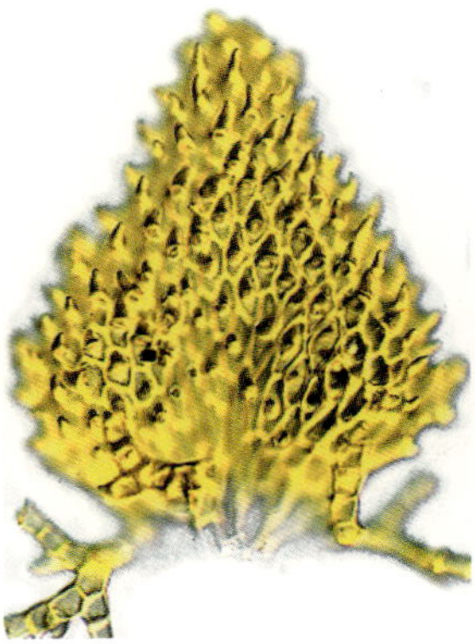

Bryonoguchia molkenboeri **spinose** papillae.
100 μm

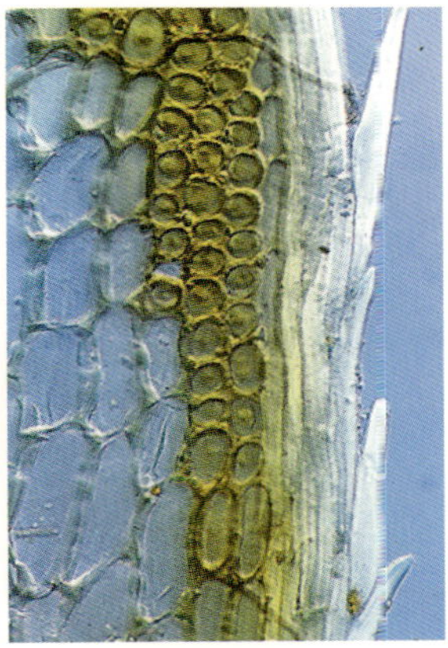

Syrrhopodon elongatus **spinose** leaf margin.
50 μm

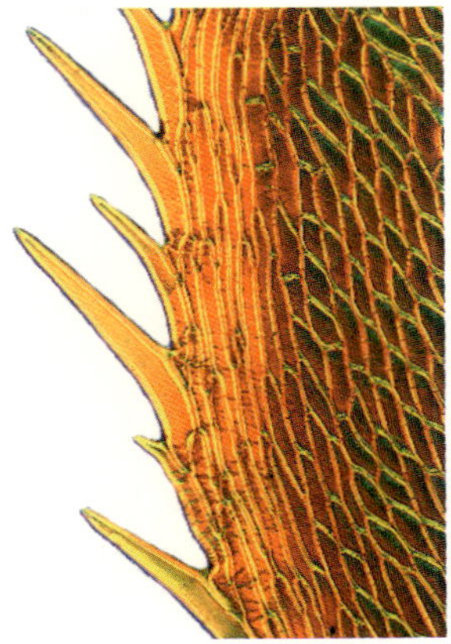

Pseudobryum speciosum **spinose** leaf margin.
100 μm

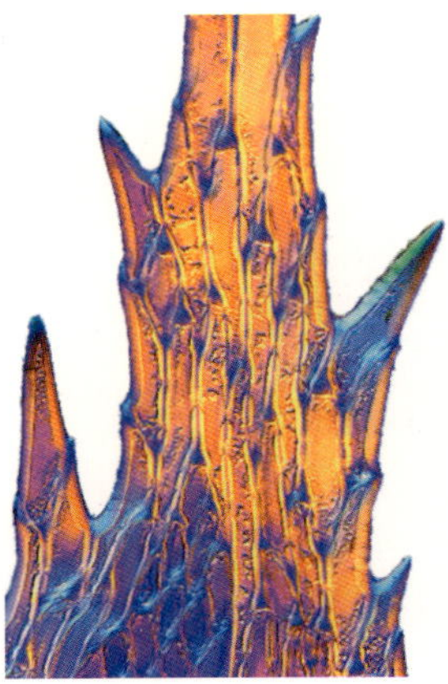

Hypopterygium commutatum **spinose** apiculus.
25 μm

Racomitrium curiosissimum **spinose** hair-point.
25 μm

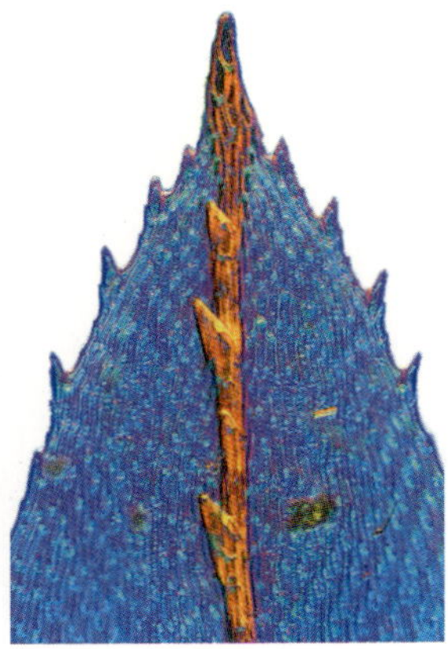

Hypnodendron spininervium **spinose** costa.
100 μm

spinulose — having tiny sharp **teeth** (•below, left).
spiral — (1, plane) resembling the shell of a snail, (2, solid) resembling a corkscrew or helix.
splash-surface — in some **thallose** liverworts, the horizontal surface of the **antheridiophore** or male **receptacle** which exploits the energy of raindrops to disperse **spermatozoids** (•below, middle).
spongiose — sponge-like in texture.
sporangium (plural **sporangia**) — the **spore-sac** portion of a **sporophyte** (•below, right).

Atrichum angustatum **spinulose** costa.
100 µm

Marchantia berteroana **splash-surface**.
1 mm

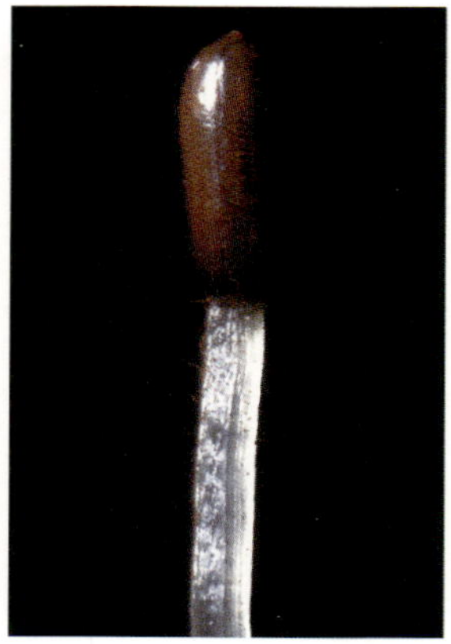
Schistochila ciliata **sporangium**.
1 mm

spore — a minute, usually **haploid**, spherical, and **unicellular** body produced in a bryophyte **sporophyte** by **meiosis** (•below).

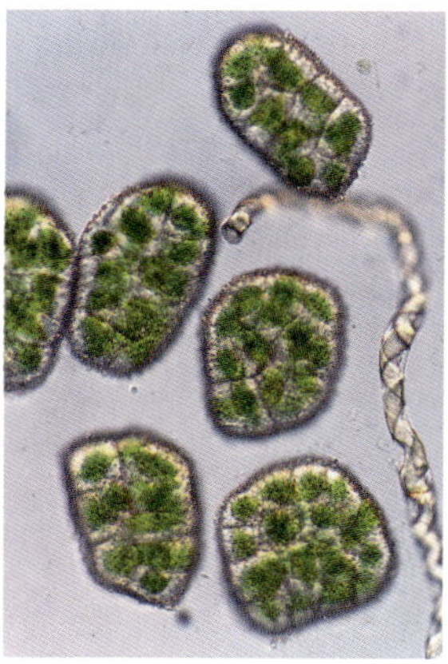
Dendroceros validus multicellular **spores**.
100 µm

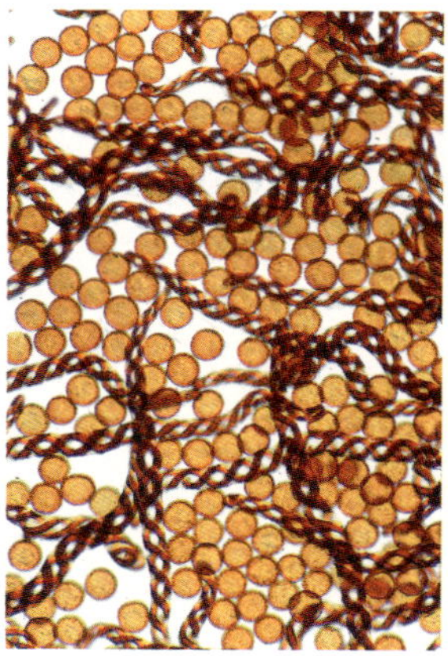
Pedinophyllum monoicum **spores** and elaters.
100 µm

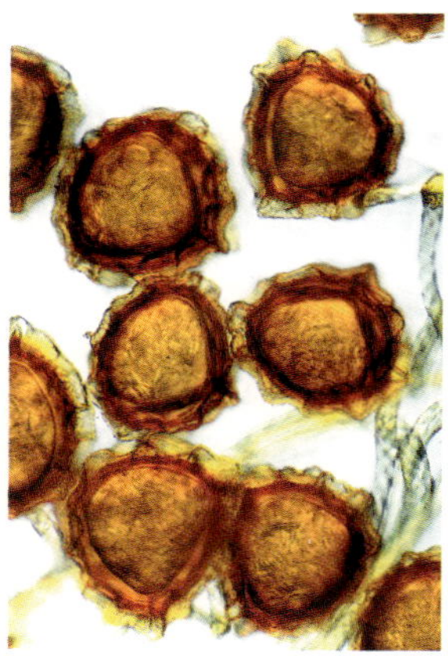
Petalophyllum australe **spores**.
50 µm

sporeling — the stage of a bryophyte's growth between **spore germination** and maturity.

spore-sac — a **spore**-containing cavity in the **capsule** that surrounds the **columella** in most mosses, but overarches it instead in species of *Sphagnum* and *Andreaea*.

sporocyte (or **spore mother cell**) — a **diploid** cell in an embryonic **sporangium** that undergoes **meiosis** and produces **haploid spores**.

sporogenous layer — the layer or mass of cells in an embryonic **sporangium** which later produces **spores** by **meiosis** (*synonym* **archesporium**).

sporophyte — the **spore**-producing generation. It typically begins as a **diploid zygote** formed by fusion of a single **sperm** (**antherozoid**) with an **egg** (**ovum**) inside an **archegonium** of a **gametophyte**. The zygote germinates and matures into a **foot** that's still embedded in the gametophyte, a stalk (**seta**), and a **capsule** (**sporangium**) which produces **haploid** spores by **meiosis**. For much of its supplies of water, mineral nutrients, and energy, it depends on the parent gametophyte.

spreading — making an angle with the stem of (1) more than 45°, (2) more than 60° (•below).

Polytrichadelphus magellanicus **spreading** leaves. 1 mm

Dicranum plurisetum **spreading** leaves. 1 mm

Macromitrium longipes **spreading** leaves. 1 mm

COMPARISON OF "SPREADING" TERMS:

appressed — flattened against the stem.
erect — parallel with the stem but not appressed.
erecto-patent — (1) less than 25°, (2) 20°–45°.
erect-spreading — about 45° or less.
patent — (1) 26°–45°, (2) 45° or more.
patulous — 46°–90°.
spreading — (1) more than 45°, (2) more than 60°.
squarrose — 90°.
squarrose-recurved — 90° with recurved tips.
wide-spreading — nearly 90°.

squalid — dingy or dirty (*synonym* **sordid**).

squamose — covered over with **scales** or made up of scales.

squamulose — covered over with minute **scales** or made up of minute scales.

squarrose — making an angle of 90° with the stem (*see* **spreading** for a comparison of similar terms).

squarrose-recurved — making an angle of 90° with the stem, with the tip curved downwards as well (*see* **spreading** for a comparison of similar terms).

staining — a general term for the vast body of techniques for coloring cell contents and cell walls that are otherwise hard to distinguish under ordinary bright-field microscopy.

stegocarpous — said of a **capsule** that opens by means of a lid (**operculum**) (*the opposite of* **astomous**, lacking a **lid**, and **cleistocarpous**, opening by the decay or breakdown of the capsule wall).

Encalypta vulgaris **stellate** leaf surface papillae. 100 μm

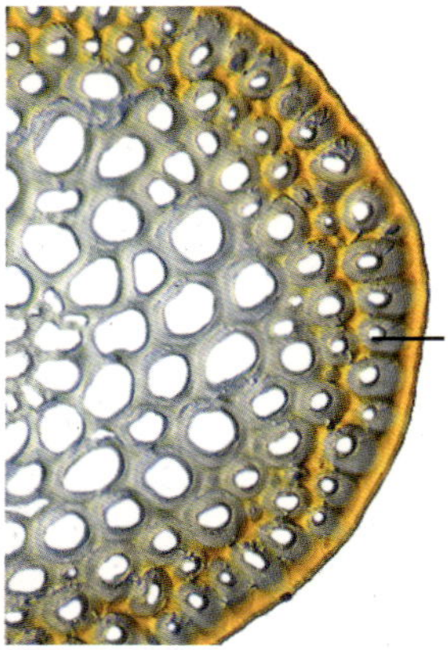

Rhaphidorrhynchium amoenum **stereids** (stem xs). 50 μm

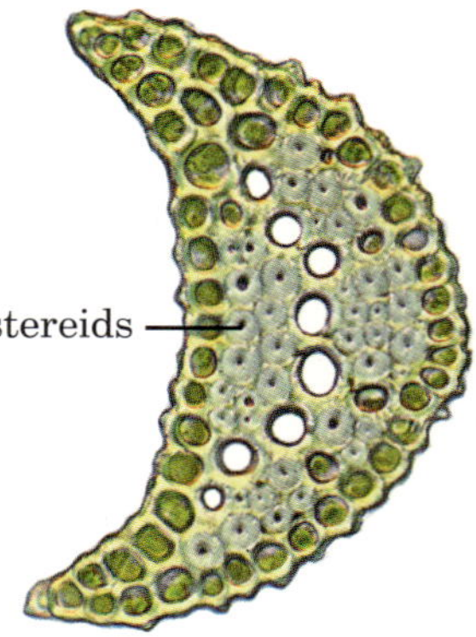

Echinodium hispidum **stereids** (leaf xs). 100 μm

stellate — star-shaped (•above, left).

stem — a main or branch **axis** bearing leaves and sex organs and growing from an **apical** cell.

stem-calyptra — in the liverworts, a hollow **fleshy** structure that's derived from both stem and **archegonial** tissue and protects the embryonic **sporophyte**. It's a type of **perigynium** and has the same function as a **perianth** and **bracts** (*synonyms* **shoot-calyptra** and **coelocaule**) (*compare with* **calyptra**, an embryo-protecting structure that's derived from only the wall of the archegonium).

stereids or **stereid cells** — "stiffener" cells that support the leaves and stems of some mosses. They're long, slender, and thick-walled (fiber-like), and occur in bundles in the **costa** and **borders** of leaves or in the **central cylinder** and toward the outside of stems (•above, middle and right).

sterome — the sum total of **stereids** in a moss plant.
stipe — the **erect**, unbranched trunk-like stem of a **dendroid** or **frondose** moss (•below, left).
stipitate — having a **stipe** or stalk (•below, left).
stolon — a slender, elongate spreading branch or stem, with small and often distinctly shaped leaves, which arches away from its parent plant like a runner (stolon) of a strawberry plant, "rooting" where it touches the **substratum** by producing **rhizoids**.
stoloniferous — producing **stolons**.
stoloniform — like a **stolon** or strawberry "runner".
stoma (plural **stomata**) — a **pore** for gas exchange, in the **sporophyte** wall (absent in liverworts, and in mosses in only the **capsule** base, the **apophysis**), usually surrounded by two **guard cells** that control its opening (•below, middle and right).
stomatose — having **stomata**.

Hypnodendron comatum **stipe**. 5 mm

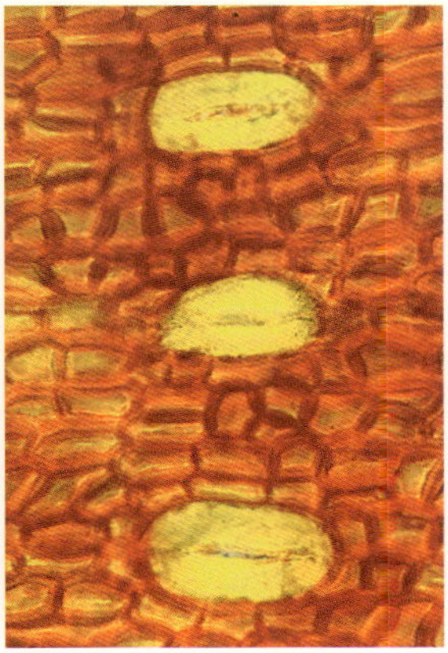

Psilopilum crispulum **stomata**. 50 µm

Anthoceros laminiferus **stomata**. 50 µm

stramineous — straw-colored.
strangulate — said of a **capsule** that's deeply **constricted** below the mouth (•below).
stratose — layered.

Dicranum fasciatum **strangulate** capsules (dry). 1 mm

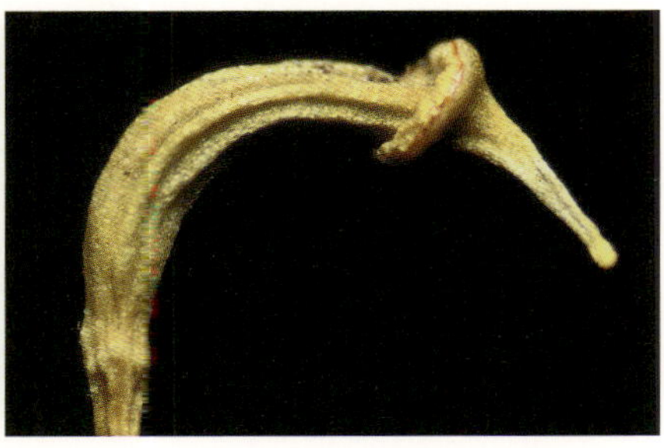

Racopilum robustum **strangulate** capsule (dry). 1 mm

stria (plural **striae**) — a longitudinal line or ridge.
striate — marked with longitudinal lines or ridges (**striae**) (•below, left and middle) (*synonym* **lineate**).
strict — straight and rigid (•below, right).
strigose — covered with rigid **appressed** hairs (*see* **hairy** for a comparison of similar terms).

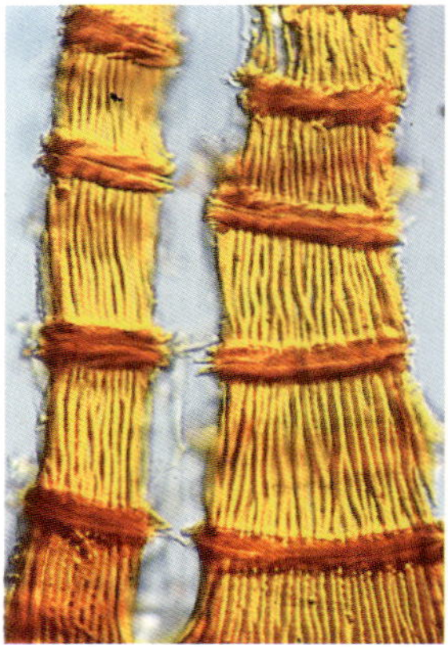

Dicranum platycaulon **striate** peristome tooth. 10 μm

Schistochila sp. **striate** hyaline setae. 1 mm

Racomitrium crispulum **strict** leaf cell rows. 50 μm

striolate — marked with fine or delicate longitudinal lines or ridges (**striolae**) (•below, left).
struma (plural **strumae**) — a goiter-like swelling on one side of the base of a structure.
strumose — having a goiter-like swelling (**struma**) on one side of a structure's base (•below, middle).
strumulose — having a small or indistinct swelling on one side of a structure's base (•below, right).

Archeophylla schusteri **striolate** leaf surface. 50 μm

Leucobryum candidum **strumose** capsules. 1 mm

Ceratodon purpureus **strumulose** capsules. 1 mm

stylus — in the leafy liverwort genus *Frullania*, a triangular flap between the **lobule** and the stem.
sub- — a prefix meaning *nearly*, *almost*, or *somewhat*, as in **subglobose** (•below) or else *under* or *below*, as in **suboral**.

Leucobryum candidum **sub**tubulose leaves. 1 mm

Breutelia pendula **sub**globose (immature) capsules. 1 mm

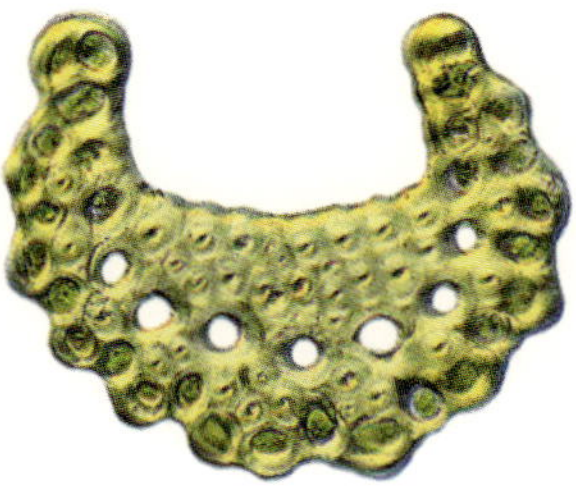

Campylopus clavatus channeled **subula** (xs). 50 µm

suberose — resembling cork in texture.
suboral — just below the mouth of a moss **capsule**.
subpercurrent — said of a **moss costa** that doesn't quite extend to the leaf **apex** (•below, left) (*see* **percurrent** for similar terms).
substratum (plural **substrata**) — whatever an organism is growing on, such as soil, rock, lignum, bark, leaves, or something man-made.
subula (plural **subulae**) — a long and slender needle- or **awl**-like point (•left).
subulate — shaped like an **awl**, **tapering** steadily to a fine point from a broad base (•below, middle and right).

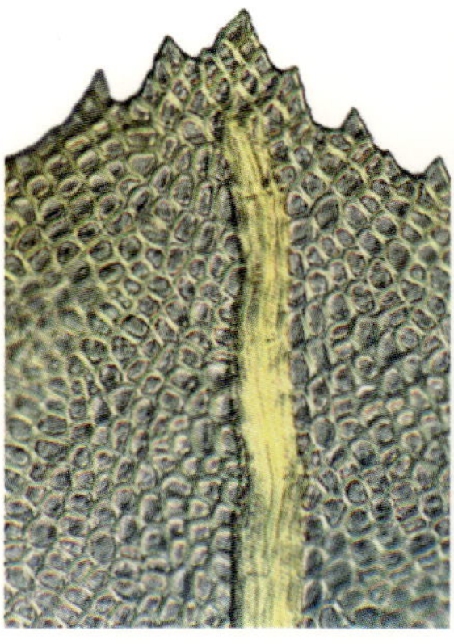

Aulacomnium heterostichum **subpercurrent** costa. 100 µm

Cryptopodium bartramioides **subulate** leaves. 5 mm

Dendroligotrichum dendroides **subulate** leaves. 10 mm

succubous — one of the three common ways that a leaf is attached to the stem in the species of leafy liverworts that have leaves more or less flattened into one plane (•below, left) (*see* **incubous leaf insertion** for a comparison of those three types).

succulent — (1) juicy, (2) having thick **fleshy** leaves, stems, or thalli (•below, middle and right) (*synonym* **fleshy**).

Marsupidium perpusillum **succubous** leaf insertion. 1 mm

Verdoornia succulenta **succulent** habit. 5 mm

Riccardia sp. **succulent** habit. 1 mm

sulcate — strongly **plicate**, with deep longitudinal grooves, furrows, or folds (•below) (*compare with* **plicate**, less strongly **grooved**, **furrowed**, or folded longitudinally (*synonym* **pleated**), and **striate** and **striolate**, weakly plicate).

Aulacomnium heterostichum **sulcate** capsules (dry). 1 mm

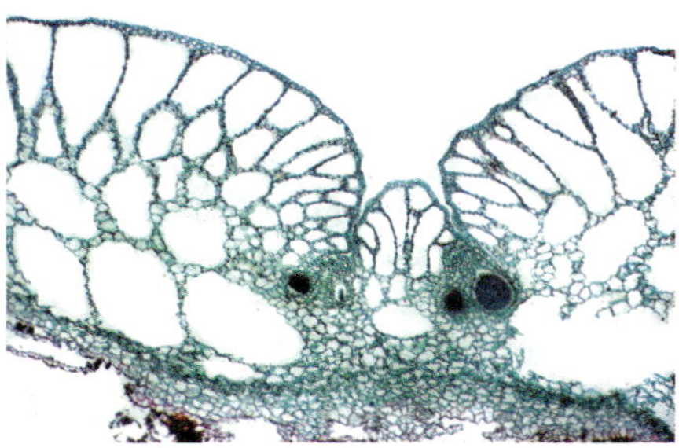

Riccia sp. **sulcate** thallus (xs). 100 µm

sulcus (plural **sulci**) — a groove or furrow (•above).

superficial — on the surface, hence used to describe **stomata** with pores that are in the same plane as the surface of the **epidermis** (*see* **cryptopore** and **phaneropore**).

surculus (plural **surculi**) — (1) a reduced branch or leaf that functions as a **vegetative propagule** (•below), (2) a sucker-like outgrowth, (3) an **erect** secondary stem growing from a horizontal primary stem or **rhizome**.

Philonotis tenuis **surculi** massed at tip of shoot. 1 mm

Philonotis tenuis **surculus** (whole mount). 0.5 mm

Pohlia ochii **surculus** (whole mount). 100 μm

suture — a junction between adjacent parts such as the **valves** of a still-unopened *Andreaea* **capsule**.

sword moss — the common name given to the moss *Bryoxiphium norvegicum*. Its crowded and strongly flattened (**distichous**) leaves resemble a sword.

sylvicolous — growing in forests (•below, left).

symbiosis — a long-term association of two or more **species**. Usually at least one of those species lives on or inside another, and typically they all belong to unrelated groups such as bacteria, fungi, algae, bryophytes, and flowering plants. Fungi are known to live inside some **dung mosses** (Splachnaceae), and perhaps boost mineral uptake. Nitrogen-fixing cyanobacteria such as *Nostoc* live inside some hornworts (•below, right) and **thallose** liverworts.

Habitat of **sylvicolous** bryophytes.

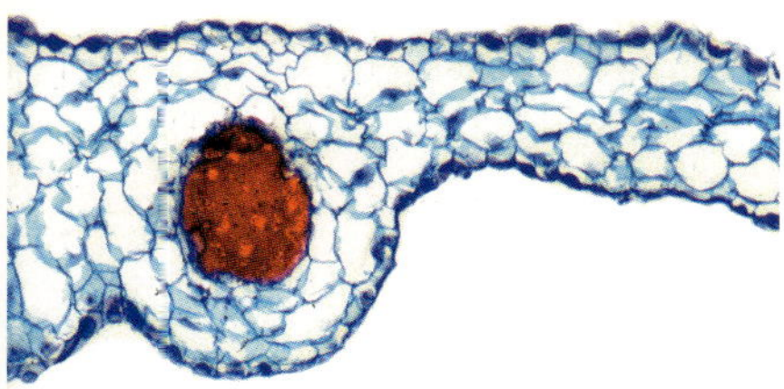

Anthoceros sp. **symbiotic** nitrogen-fixer (xs). 100 μm

sympatric (noun **sympatry**) — said of **taxa** inhabiting the same or overlapping regions (*the opposite of* **allopatric**, inhabiting separate regions).

symplastic water — water held within the cell protoplast or **lumen**. *Sphagnum* leaves can hold up to 25 times their dry weight because most of their bulk is dead **porose** cells that readily fill with symplastic water (•below, left). Also, the **stereids** of some mosses which lack a **central** (axial) **strand** can conduct water symplastically (*compare with* **apoplastic water**, water that's held in the polymers making up the cell walls, and **capillary water**, water held externally by surface tension).

sympodial — branching repeatedly with stems of limited growth which are continually replaced by side branches (**innovations**).

synoicous — **monoicous**, producing both male and female sex organs (**antheridia** and **archegonia**) in the same **inflorescence** and not grouped or clustered by sex (*synonym* **androgynous**).

systematics — the science of identifying, naming, and classifying organisms (*synonym* **taxonomy**).

systylious — said of a moss **capsule** in which the **operculum** remains attached to the tip of the **columella** after the capsule has opened.

tannin (or **tannic acid**) — a complex polymer made by some plants and used as tanning compounds, as mordants in dyeing, and as medical astringents. The yellow to brown pigments in the cell walls of the leaves of many species of *Andreaea* mosses are tannins (•below, right).

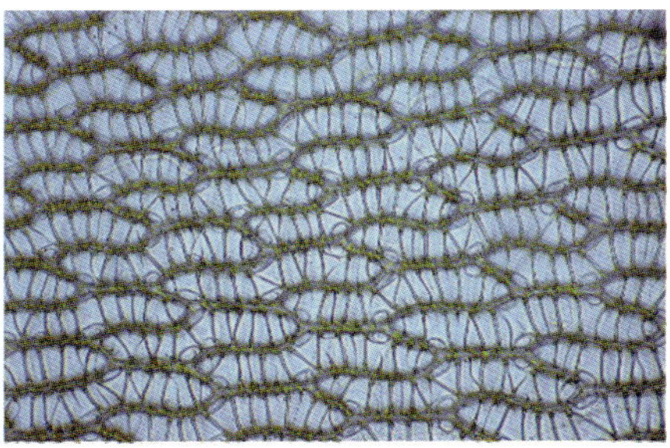

Sphagnum cristatum leaf storing **symplastic** water. 100 μm

Andreaea mutabilis **tannin**-pigmented leaf cells. 50 μm

tapering — narrowing toward one end.

tapetum (plural **tapeta**) — a layer of nutritive cells in a **sporangium**, sometimes called **nurse cells**.

taphrophilous — thriving in ditches or trenches.

tartareous — having a thick, rough, and crumbling surface texture.

taxon (plural **taxa**) — a group of organisms that's been given a name according to nomenclatural rules. The form and especially the suffix of the name depend on the rank the taxon has been assigned, and so can vary in different classifications.

taxonomy — the science of identifying, naming, and classifying organisms (*synonym* **systematics**).

teeth (singular **tooth**) — (1) the **serrations** of a leaf **margin** or other structure (•below, left and middle), (2) in mosses, the divisions of either a **single peristome** (•below, right) or of the **exostome** of a **double peristome** (divisions of the **endostome** are called **segments** or **processes**).

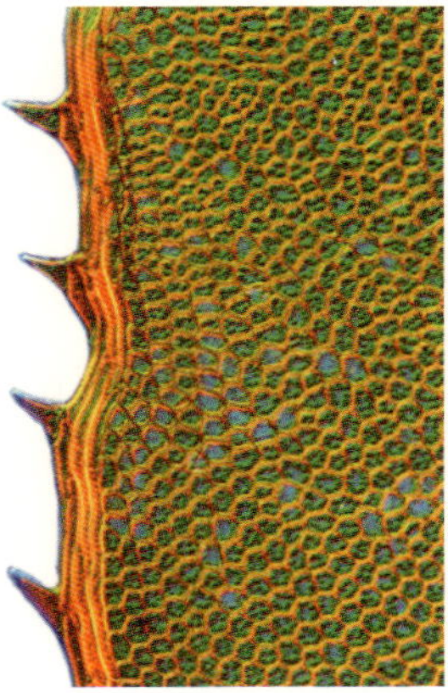

Plagiomnium cuspidatum **teeth** of leaf margin.
100 µm

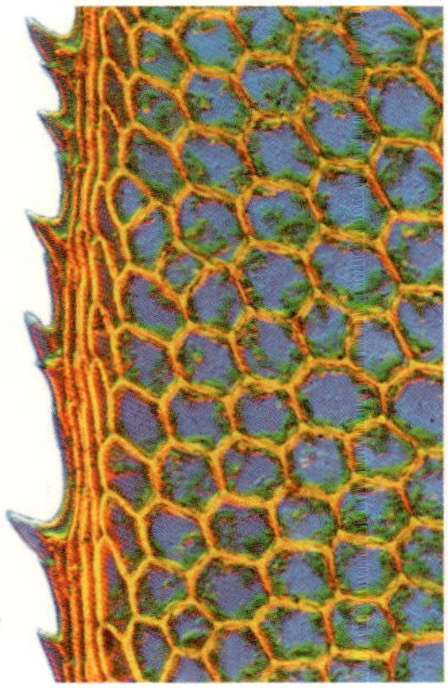

Calyptrochaeta cristata **teeth** of leaf margin
100 µm

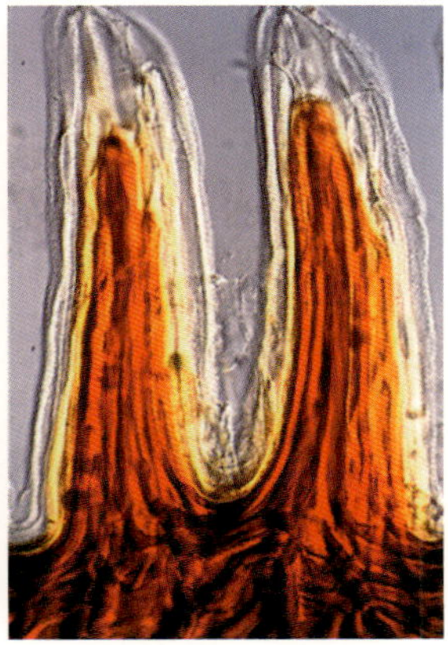

Pogonatum subulatum peristome **teeth**.
100 µm

teniola (plural **teniolae**) — **intramarginal border** of **linear hyaline** cells (•below, left).

terete — cylindrical in form and therefore round in cross-section (•below, right).

Calyptopogon mnioides **teniola**.
100 µm

Pyrrhobryum bifarium **terete** seta xs.
50 µm

terminal — at the end, tip, or **apex** of a structure (•below) (*the opposite of* **proximal** or **basal**).

terminal branch — in leafy liverworts, a branch produced by the **apical** cell rather than by cells further down the stem.

Jamesoniella colorata **terminal** perianths.
1 mm

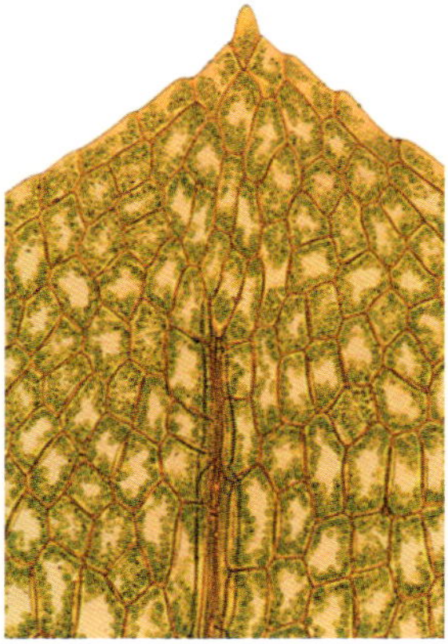

Bryobeckettia bartlettii **terminal** apiculus.
100 µm

Tetraphidopsis pusilla **terminal** gemmae.
1 mm

ternate — said of three structures that are arranged around an **axis**.

terpenoids — a group of chemical compounds formed from isoprene units. They give many liverworts a distinctive odor and can cause rashes in humans.

terricolous (or **terrestrial**) — growing on or preferring a soil **habitat** (•below, left and middle).

tessellated — resembling a mosaic of square tiles (•frontispiece and below, right).

Neohodgsonia mirabilis **terricolous** species.
1 mm

Chiloscyphus semiteres **terricolous** habit.
5 mm

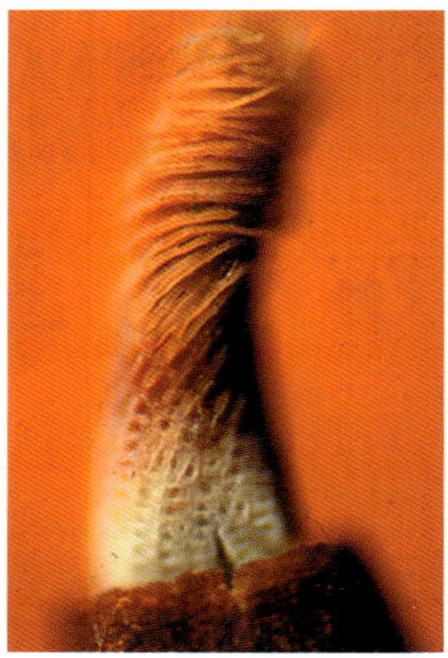

Tortula muralis **tessellated** peristome base.
100 µm

tetrad — a group of four developing **spores** formed by the **meiotic** division of a **spore mother cell**. Because the spores are tightly pressed together in the tetrad, the **proximal** face of each one is deformed by the other three spores (*see* **triradiate mark** and **triradiate ridge**). As a result of their tardy separation from the tetrad, the mature spores can be ellipsoidal, angular, or kidney-shaped (*see* **trilete spores**).

tetragonal — four-angled.

tetrahedral apical cell — the four-faced cell at the tip of a bryophyte shoot which looks like an upside-down pyramid and divides on its three rear faces to form the plant's tissues (*see* **cutting face**).

tetraploid — an organism that has four full sets of **chromosomes** in each of its nuclei.

thallose (or **thalloid**) — made up of a flat plate or sheet of tissue. All hornwort **gametophytes** are thallose, whereas those of liverworts are thallose or leafy or intermediate between the two (•below, right), and those of all but a few mosses are leafy.

thallus (plural **thalli**) — a plant body that's a flat mass of tissue.

Cheilolejeunea albovirens **tiered** seta cells.
100 µm

Balantiopsis tumida **tiered** leaf cells.
100 µm

Monoclea forsteri **thallose** liverwort.
1 cm

theca (plural **thecae**) — in mosses, the main body (**urn**) of a **capsule**

thecal — (1) said of branches that are produced below the tip of the stem (*the opposite of* **athecal**, produced at the tip of the stem), (2) referring to the **capsule** of a moss or liverwort.

thin-layer chromatography (**TLC**) — a technique for separating compounds by means of their different solubility in mixtures of solvents.

tiered — arranged in horizontal or **transverse** rows (•above, left and middle) (*compare with* **seriate**, arranged in rows).

tissue — any sheet or mass of cells making up an organ.

tmema (plural **tmemata**) — an abscission cell at the base of a **gemma** (•below, left).

tomentum (plural **tomenta**) — a dense, woolly felted covering made up of long **rhizoids** (adjective **tomentose**) (•below, middle and right).

tooth — (*see* **teeth**).

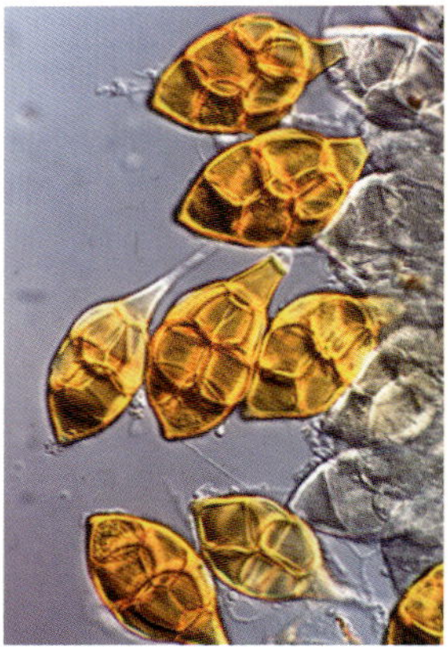

Aulacomnium androgynum **tmemata**.
50 µm

Philonotis tenuis shoots with basal **tomentum**.
1 mm

Breutelia pendula shoots with **tomentum**.
1 mm

torquate — twisted, spiraled, or rotated around a central **axis** (•below) (*compare with* **dextrorse**, twisted like a standard right-hand thread (•below, right), and **sinistrorse**, twisted like a left-hand thread).

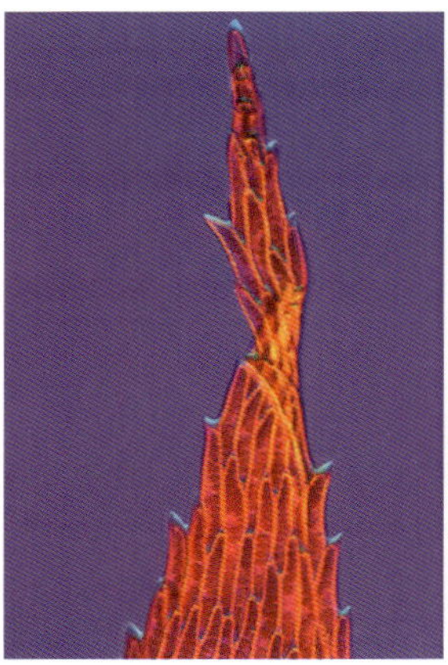

Rhynchostegium serrulatum **torquate** leaf apex.
100 µm

Tortula muralis **torquate** peristome.
1 mm

Racopilum convolutaceum **torquate** dextrorse seta.
1 mm

tortuose — irregularly bent or twisted (•below) (*similar to* definition #2 of **contorted**).

totipotency — the ability of a fragment of any part of a plant to grow into a new plant. Most bryophytes are totipotent. However, in mosses the fragments don't send up new shoots directly, but instead sprout **protonemata**, which in turn develop **buds** and then shoots.

Atrichum angustatum **tortuose** leaves (dry) at shoot apex. 0.5 mm

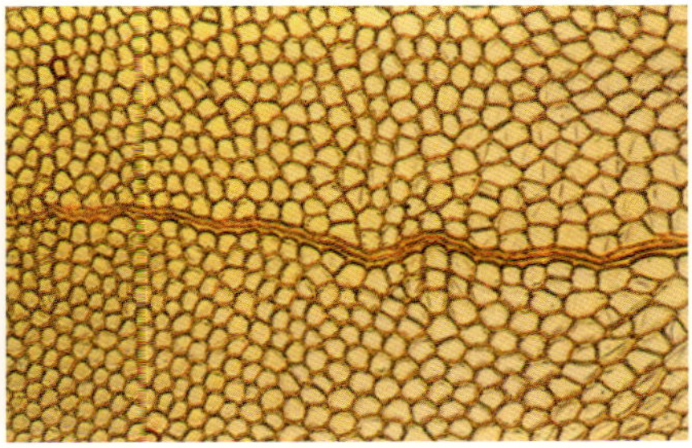

Distichophyllum pulchellum **tortuose** costa. 100 μm

trabeculae (singular **trabecula**) — (1) cross-bars or projections on the back (inner surface) of a tooth of the **outer peristome** (**exostome**) (•below, left), (2) slender supportive strands that support **air-chambers** in many marchantialean liverworts (•below, middle).

trabeculate — having cross-bars or **trabeculae** (•below, left and middle).

transfer cells — specialized cells at the junction of the **gametophyte** and **sporophyte** that function in nourishing the sporophyte (•below, right).

Funaria hygrometrica **trabeculate** peristome. 100 μm

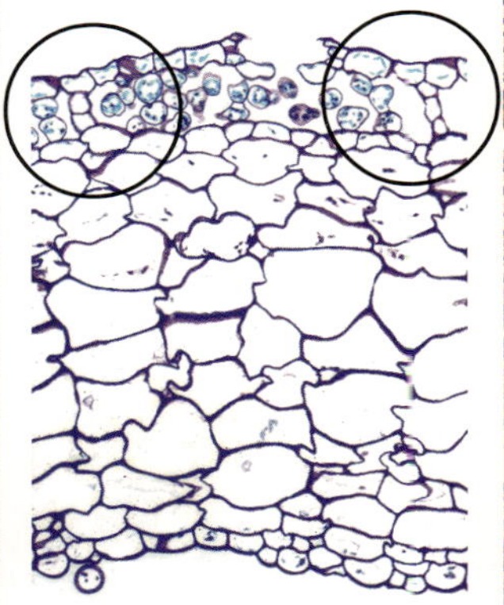

Marchantia sp. thallus xs, air-chamber **trabeculae**. 100 μm

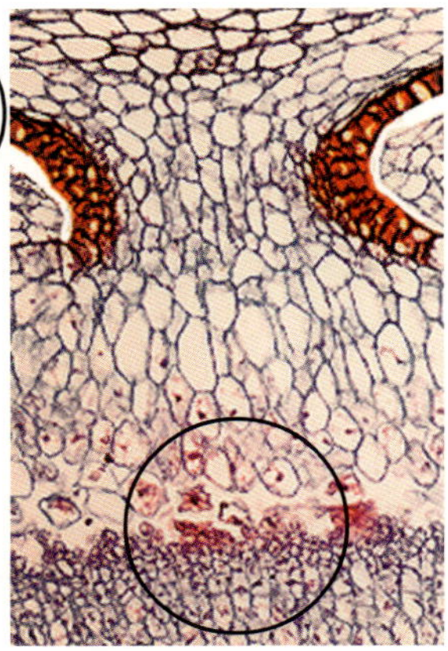

Sphagnum sp. **transfer cells**. 100 μm

transverse — (1) at right angles to the **axis** or the middle of a structure such as a stem or leaf (•below) (*compare with* **longitudinal**, parallel to the axis or the middle of a structure), (2) one of the three common ways that a leaf is attached to the stem in the species of leafy liverworts which have leaves that are more-or-less flattened into one plane (•opposite page) (*see* **incubous leaf insertion** for a comparison of those three types).

Neckera pennata **transverse** leaf undulations. 1 mm

Ulota lutea **transverse** wall thickenings. 100 µm

Racomitrium crispulum thin **transverse** walls. 50 µm

trapezoid or **trapeziform** — (1) a shape with four sides, none of which is parallel to any other, (2) a shape with four sides, two of which are parallel but of different lengths.

travertine — a porous brown limestone widely used as a decorative building stone and formed from **tufa**, calcium carbonate that's been accreted by **calcicolous** mosses (*see* **didymodontolith**).

tri- — a prefix meaning *three*.

triangular — in the shape of a triangle (•below).

Philonotis tenuis **triangular** leaves. 1 mm

Schistidium apocarpum **triangular** peristome teeth. 1 mm

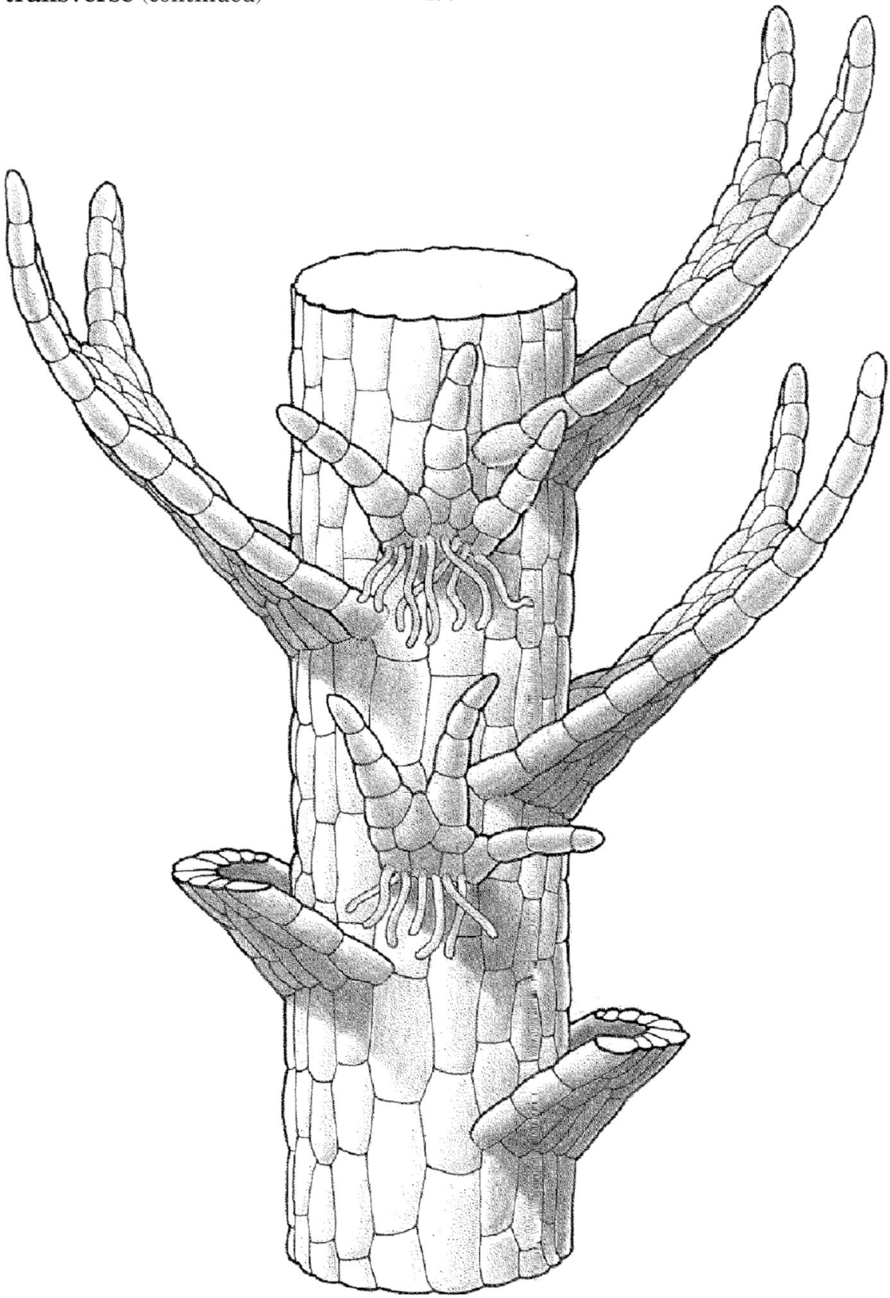

Transverse leaf insertion in leafy liverwort.

trichotomous — regularly dividing into three roughly equal branches or parts.
tridentate — having three **teeth**.
trifarious (or **tristichous**) — arranged in three rows or **ranks** (*see* **phyllotaxy**).
trifid — clearly divided into three segments, parts, or **lobes**.
trigone — a triangle-shaped wall thickening in the corner of a cell where it abuts against two other cells (•below, left and middle). The presence, shape, and size of trigones are characteristic of some liverwort genera, and so are useful tools for identification and classification.
trigonous — (1) having three corners, (2) **triangular** in cross-section.

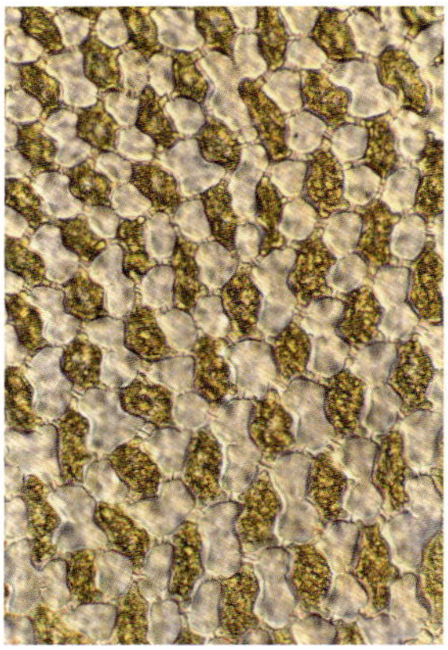

Chandonanthus squarrosus massive **trigones**. 50 µm

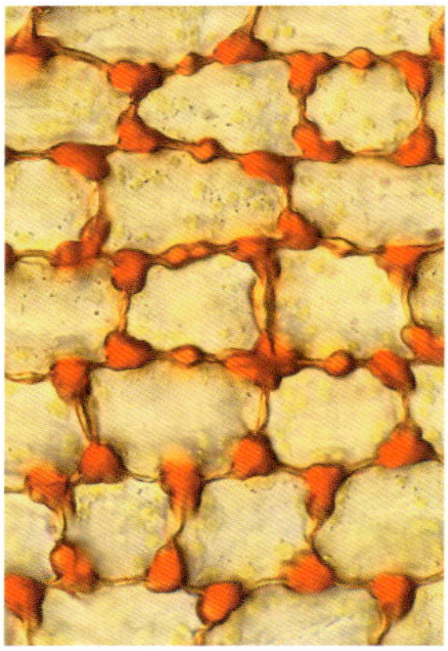

Rhaphidorrhynchium amoenum capsule wall **trigones**. 25 µm

Phaeoceros hirticalyx **triradiate ridges**. 50 µm

trilete spore — a **spore** that has a three-pronged (**triradiate**) scar on its wall where it was in contact with three other spores of a **tetrad** (*compare with* **monolete spore**).
triquetrous — having three angles.
triradiate — having three lines that radiate from a central point (*see* **trilete spore**).
triradiate mark — the deformation of the **proximal** face of a **spore** as a result of its being tightly pressed against the three other spores formed with it in a **tetrad**.
triradiate ridge — a **triradiate** thickening formed on the **proximal** wall of a **spore** while it's tightly pressed against the three other spores formed with it in a **tetrad** (•above, right).
triseriate — arranged in three rows.
tristichous (or **trifarious**) — arranged in three rows or **ranks** (*see* **phyllotaxy**).
tristratose — made up of three layers.

truncate — cut off **abruptly** and often squarely near the **apex** (•below).

TS — cross (**transverse**) section (*synonym* **xs**).

tubate (or **tubiform**) — shaped like a trumpet, hollow and flared at one end.

Neckeropsis calcicola
truncate stem leaf.
0.5 mm

Leucodon sapporensis
truncate leaf apex.
100 μm

Fontinalis antipyretica
truncate endostome.
0.5 mm

tuber — (1) in mosses, a **gemma**-like **brood body**, non-green and borne on **rhizoids**, usually underground (•below), (2) in liverworts and hornworts, a **propagule** consisting of a dormant **apical** cell surrounded by several protective layers of dead cells.

Bryum billardierei **tubers**.
1 mm

Bryum sp. **tubers**.
100 μm

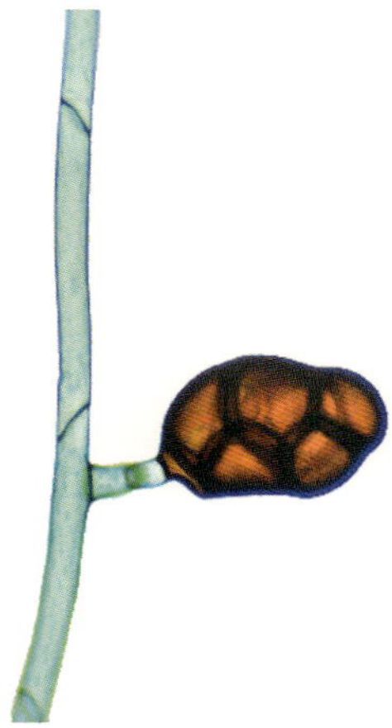

Bryum sauteri **tuber**.
100 μm

tuberculate — covered with **wart**- or peg-like projections (•below, left).
tubercule — a **wart**- or peg-like projection.
tubiform (or **tubate**) — shaped like a trumpet, hollow and flared at one end.
tubular (or **tubulose**) — having the form of a tube, usually formed by the strongly **inrolled margins** of leaves or **bracts** (•below, middle).
tufa — deposits of calcium carbonate built up around the **calcicolous** moss *Didymodon tophaceus*. The water the moss grows in is saturated with calcium bicarbonate, which precipitates out as carbonate as the moss removes carbon dioxide from the water (*see* **didymodontolith**).

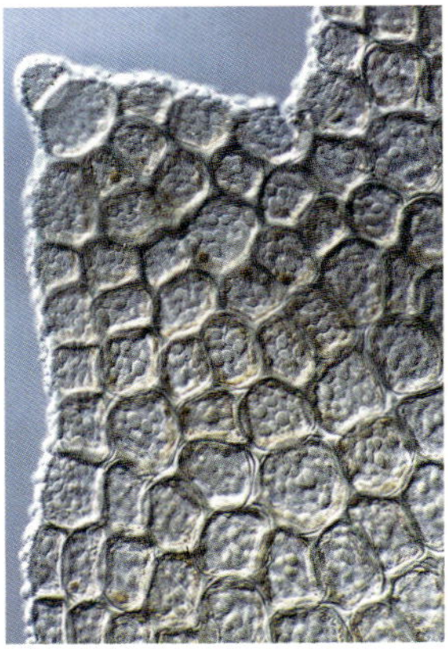

Chloranthelia berggrenii **tuberculate** leaf surface. 50 µm

Ditrichum difficile **tubular** shoot apex (xs). 50 µm

Dawsonia superba **tuft** of peristome hairs. 5 mm

tuft — (1) a clump of more or less **erect** shoots, (2) a clump or bundle of similar structures such as leaves or hairs (•above, right and below, right).

Syntrichia princeps **tufted** habit. 1 cm

Macromitrium sp. **tuft** (cutaway view). 1 mm

tufted — in clusters of **erect** but radiating **shoots** (•above, left).
tumid — swollen or **inflated**.

turbinate — shaped like a child's top, an inverted cone, or an upside-down pear.

turf — a growth form of extensive patches of crowded **erect** shoots (•below, left).

Tortula truncata **turf** (cutaway view). 10 mm

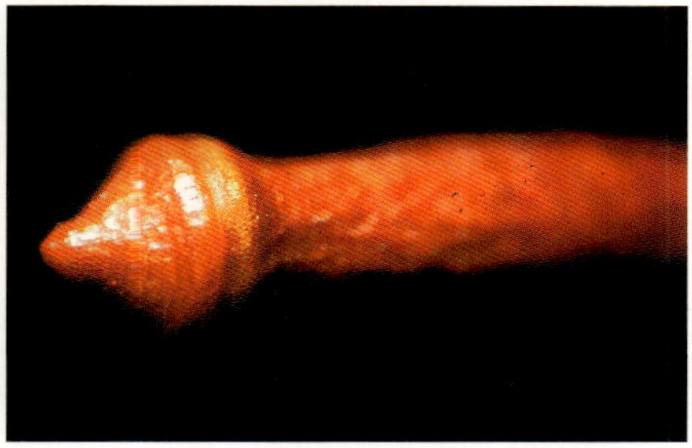

Bryum billardierei **umbonate** operculum. 0.5 mm

turgid — plump or swollen.

turgor pressure — the pressure exerted on the cell walls by the fluid within, resulting from the absorption of water into cell vacuoles.

umbilicate — shaped like a plate or shield supported by a central stalk (*synonym* **peltate**).

umbonate — **convex** but with an **abrupt**, rounded central point (•above, right).

uncinate — shaped like a hook (•below, left) (*synonyms* **aduncous**, **hamate**, and **hooked**).

underleaf — a leaf on the **ventral** stem surface, typically smaller and with a shape different from that of other leaves (•below, middle) (*synonyms* in leafy liverworts **amphigastrium** and **ventral leaf**).

underleaf ratio — the number of **underleaves** per *pair* of **lateral leaves** (•below, right).

Hypnum lindbergii **uncinate** shoot tip. 1 mm

Cyathophorum bulbosum **underleaves**. 1 mm

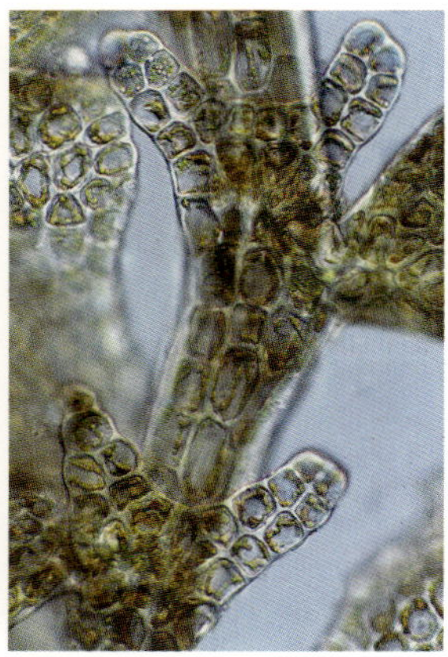

Diplasiolejeunea plicatiloba **underleaf ratio** 2:1. 100 μm

undulate — somewhat wavy (•below) (*compare with* **rugose**, strongly wavy).

Neckera pennata **undulate** leaves.
1 mm

Atrichum androgynum **undulate** leaves.
5 mm

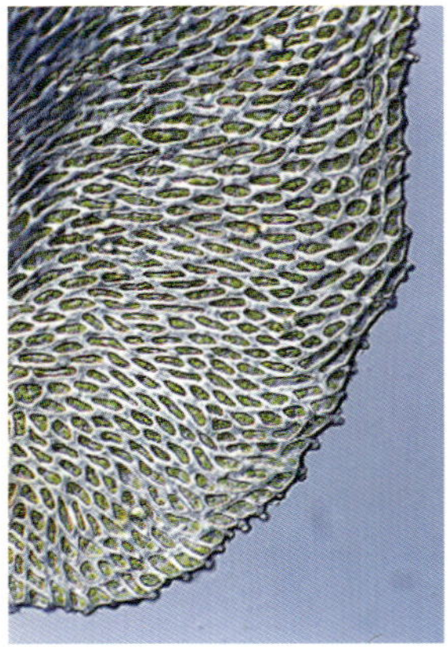

Papillaria crocea **undulate** leaf margin.
10 µm

unequal — **asymmetric** (•below, left and middle).

unguiculate — ending in a point resembling a claw (•below, right).

Rhizogonium distichum **unequal** leaf halves.
1 mm

Anomodon giraldii cell walls **unequally** thick.
10 µm

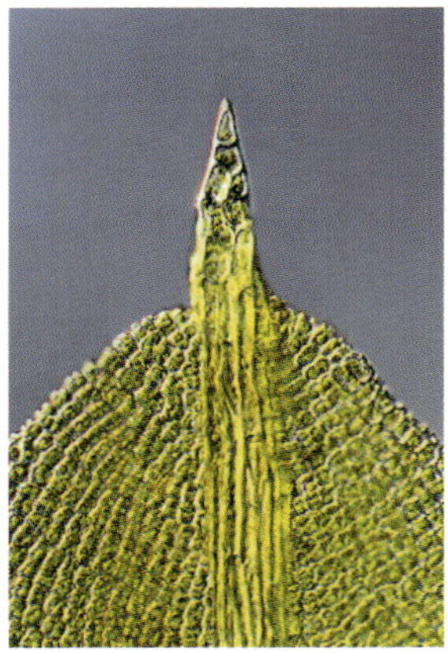

Barbula unguiculata **unguiculate** leaf apex.
100 µm

uni- — a prefix meaning *one*, as in **unipapillose** or **unistratose** (•below and next page).

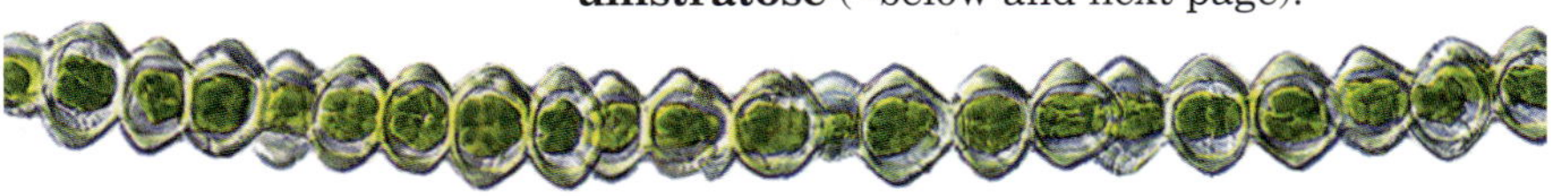

Racopilum convolutaceum **unipapillose** leaf cells (xs). 25 µm

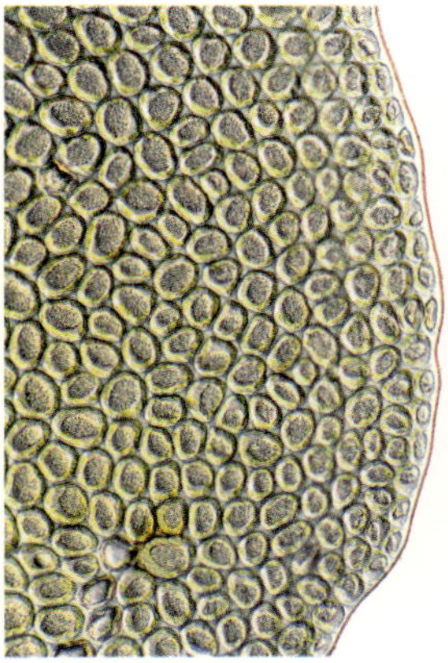

Radula silvosa one oil-body per cell.
100 µm

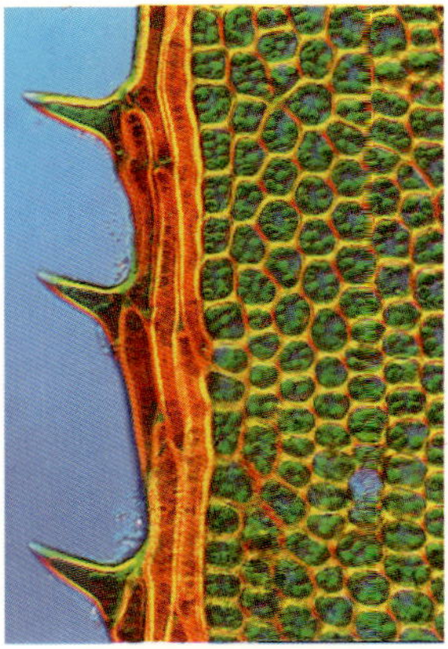

Plagiomnium cuspidatum **unicellular** teeth.
100 µm

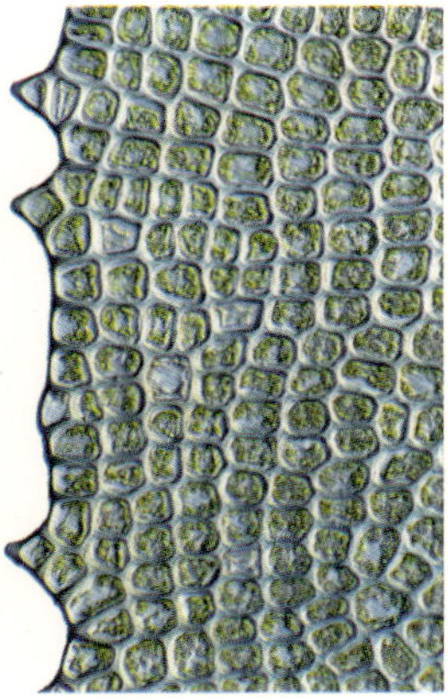

Adelanthus falcatus **unicellular** marginal teeth.
100 µm

unicellular — one-celled (• above, middle and right).
unilateral — one-sided.
unipapillose — having one **papilla** per cell (• below, left and middle, and facing page, bottom).

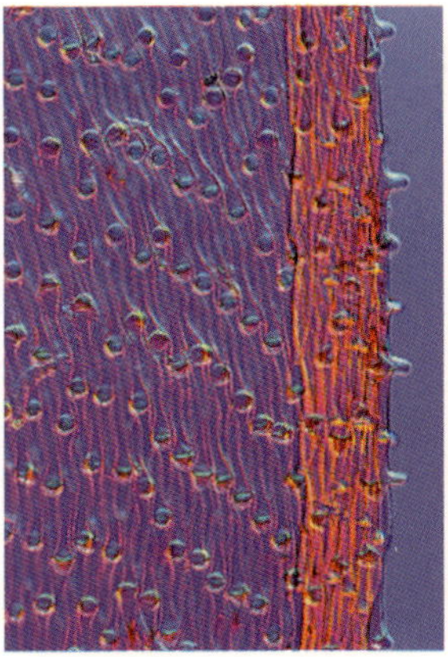

Rhaphidostichum boschii **unipapillose** leaf cells.
100 µm

Chiloscyphus mittenianus **unipapillose** leaf cells.
100 µm

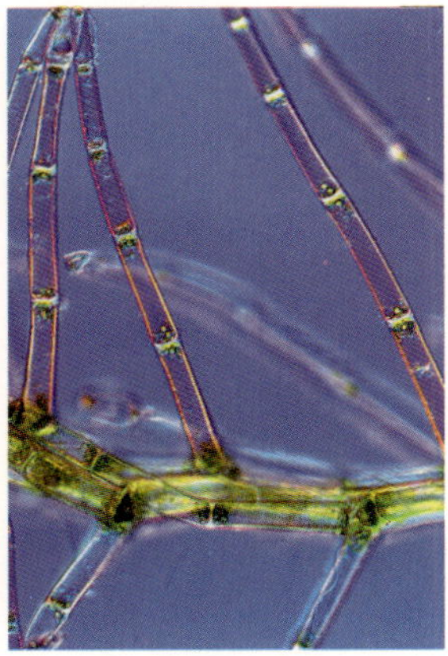

Telaranea herzogii **uniseriate** leaves.
100 µm

uniseriate — arranged in one row (• above, right).
unistratose — one cell layer thick (• below).

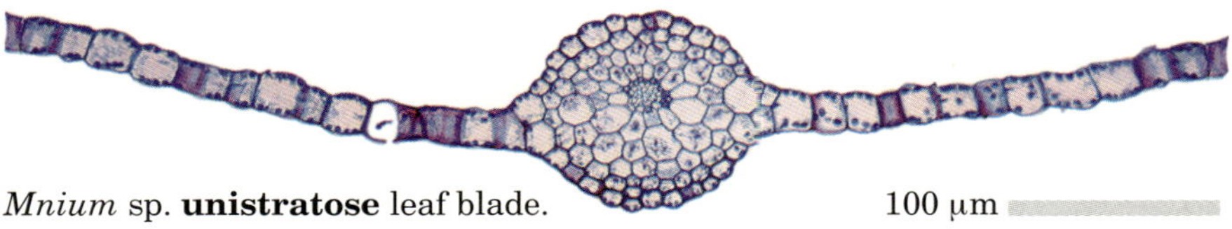

Mnium sp. **unistratose** leaf blade. 100 µm

upper-leaf — the outer (**distal**) third of a leaf.

urban mosses — mosses notably abundant in cities and other human environments (•below).

Didymodon australasiae **urban moss** in paving-stone cracks. 50 mm

Pohlia tenuifolia **urban moss** on car-wash brush. 50 mm

Bryum argenteum **urban moss** in sculptured curbing. 50 mm

urceolate — **urn**-shaped, narrowed somewhat just below the mouth and again strongly at the base (•below, left and middle) (*compare with* **cupulate** and **cyathiform**, cup-shaped with a mouth that's only slightly flared, and **campanulate**, bell-shaped with a mouth that's widely flared).

urn — in mosses, the **spore**-bearing portion of a **capsule** (•below, right) (*synonym* **theca**).

utricle — a vesicle or bladder-like structure.

Polytrichadelphus magellanicus **urceolate** capsules. 1 mm

Fissidens sp. **urceolate** capsules. 1 mm

Cyathophorum bulbosum **urn**. 1 mm

vacuole — a cavity within the cytoplasm filled with watery cell sap.

vaginant — *see* **vaginate**.

vaginant lamina (plural **laminae**) — in species of the moss genus *Fissidens*, the folded basal part of the leaf that clasps the stem (•below).

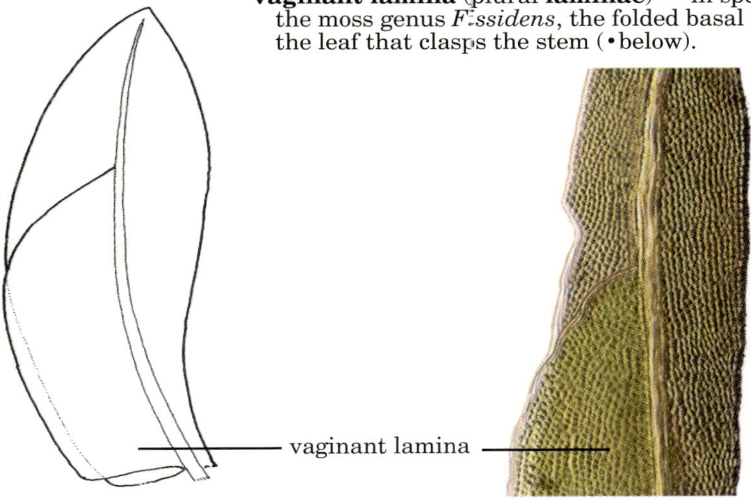

vaginate (or **vaginant**) — **sheathing** (•below, left).

vaginula — in mosses, the ring or **sheath** enveloping the base of the **seta**, derived from the **venter** of the **archegonium** after loss of the **calyptra**.

valve — in most liverworts and hornworts and the andreaeopsid mosses, the segments or flaps that the **sporangium jacket** separates into when it opens (**dehisces**) along lines of weakness (**sutures**) (•below, middle and right).

Bartramia pomiformis **vaginate** leaf bases. 1 mm

Andreaea acutifolia ssp. *acuminata* **valves**. 1 mm

Schistochila sp. **valves** of dehisced sporangium. 1 mm

vector — anything that transports **diaspores**.

vegetative — (1) non-sexual growth by **mitosis** only, (2) said of all leaves except those surrounding the sexual organs or a **seta** base (•below, left).

velutinous — velvety (*see* **hairy** for a comparison of similar terms).

venter — the swollen **basal** portion of an **archegonium**, containing a single **egg** (•below, middle). Cells at its base transport sugars from the parent **gametophyte** to the developing **sporophyte**.

ventral — said of the lower surface of a **prostrate** stem, the inner surface of a **peristome tooth**, the upper surface of a leaf (•below, right) and the lower surface of a **thallose** liverwort (*the opposite of* **dorsal**).

Rhizogonium distichum **vegetative** shoots.
10 mm

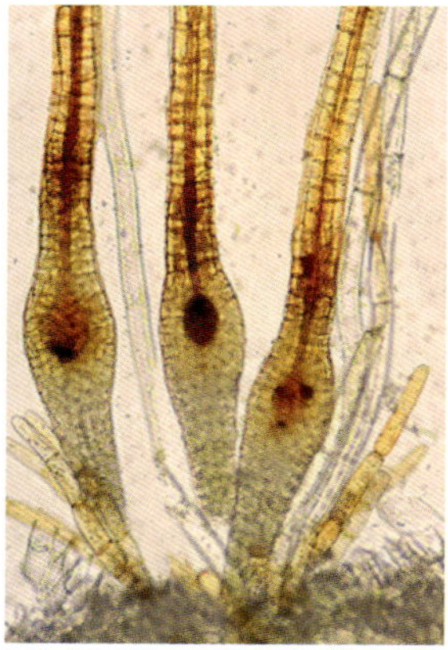

Leptostomum macrocarpon **venters**.
50 μm

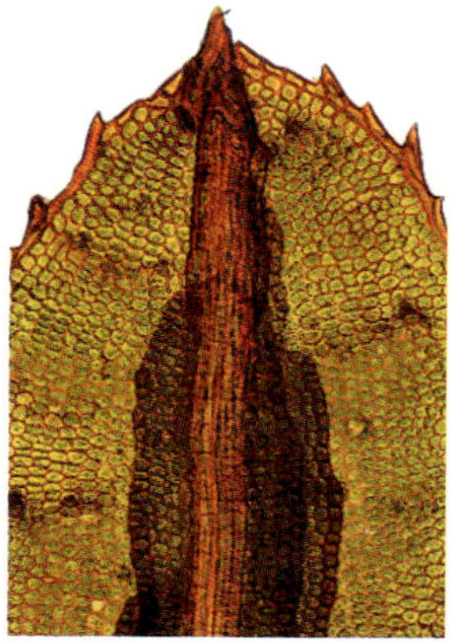

Atrichum angustatum **ventral** lamellae.
100 μm

ventral-intercalary — (1) in leafy liverworts, said of branches that are produced below the tip in the **axils** of **underleaves**, (2) any **ventral** branch located well below the **apex**.

ventral lamina — (*see* facing page).

ventral leaf — in some leafy liverworts, a modified leaf on the ventral side of the stem, usually smaller than the **lateral leaves** and with a different shape and **margin** (*synonyms* **amphigastrium** and **underleaf**).

ventral scale — in some **thallose** liverworts, a membranous **unistratose scale** on the underside of the **thallus**, often arranged with numerous other scales in two or more rows.

ventricose — shaped like a stomach, bulging or swollen on one side toward the base (*similar to* **gasteropodous**) (*compare with* **gibbous**, bulging or swollen on one side toward the top).

verdigris — blue-green, the color of strongly weathered copper (*similar to* **aeruginose**).

ventral lamina (*synonym* **apical lamina**) — the upper non-folded portion of a *Fissidens* leaf beyond the basal folded portion (which is called the **vaginant lamina**). A *Fissidens* leaf is complex, and so not surprisingly various authors have interpreted its structure in conflicting ways. The structure and its variants can easily be visualized by making large paper models of them. The diagrams below show how to do that. First, using a photo-copier, enlarge the two sets of full- and half-leaves labeled **a** and **b**. Cut everything out, and crease the two full-leaves lightly along their midribs, then glue a half-leaf to the back of each crease. Lastly, for the leaf on sheet **b**, tape to the longer side of the leaf the top of the "loose" flap on the shorter side of the leaf. In both of these leaf variants, the folded portion at the base of the leaf is called the vaginant lamina—it **sheaths** the stem, and is thought by most authors to be the true leaf. The half-leaf is called the **dorsal lamina**, and is usually interpreted as a **wing**.

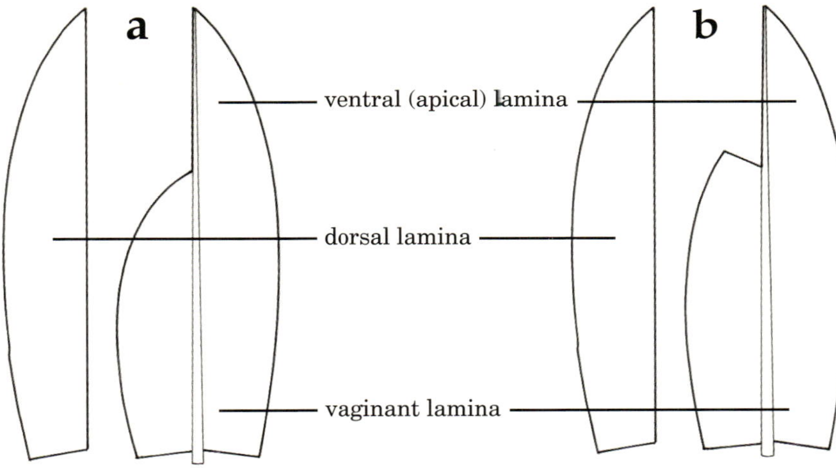

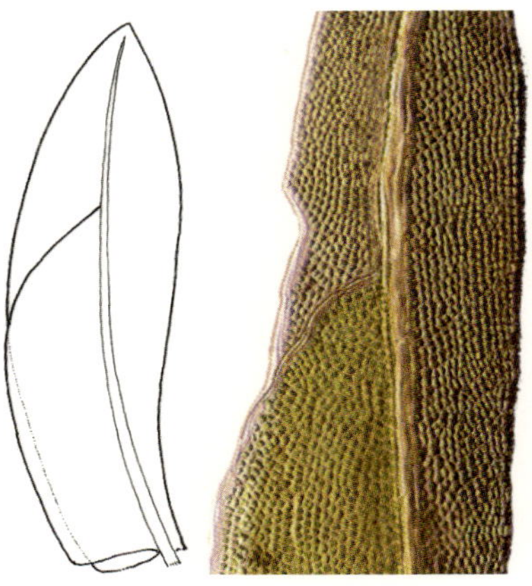

Fissidens blechnoides diagram of leaf (left) and portion of leaf. 100 µm

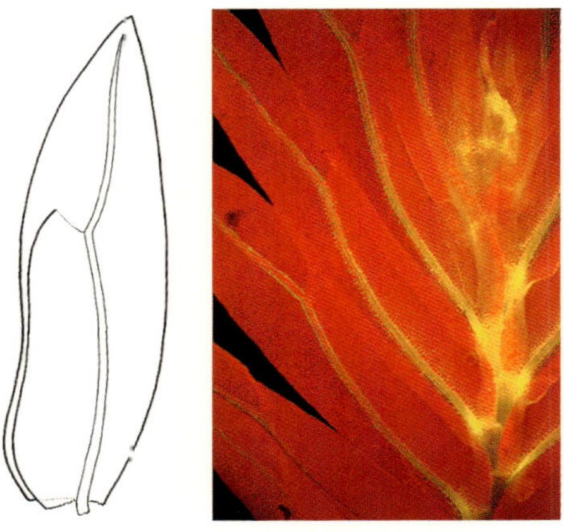

Fissidens pallidus diagram of leaf (left) and portion of shoot (ultraviolet microscope view). 1 mm

vermiform (or **vermicular**) — shaped like a worm, long, narrow, and wavy (•below, left and middle).
verruca (plural **verrucae**) — a **wart**-like swelling.

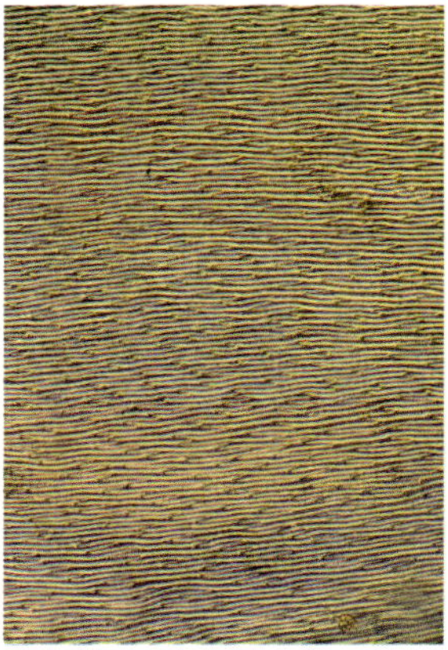

Calliergoniella cuspidata **vermiform** leaf cells.
100 µm

Hypnum lindbergii **vermiform** leaf cells.
100 µm

Temnoma quadripartitum **verrucose** leaf surface.
100 µm

verrucose — **warty** or roughened (•above, right and below, left).
verruculose — delicately or irregularly roughened.
verticillate — **whorled** (•below, middle).
vesicular — made up of blisters.
vesiculose — **inflated** like a bladder.
vestigial — said of a structure or a function that's reduced to only a remnant, trace, or hint.
viability — the ability to **germinate**.
viatical — growing alongside tracks or roads (•below, right).

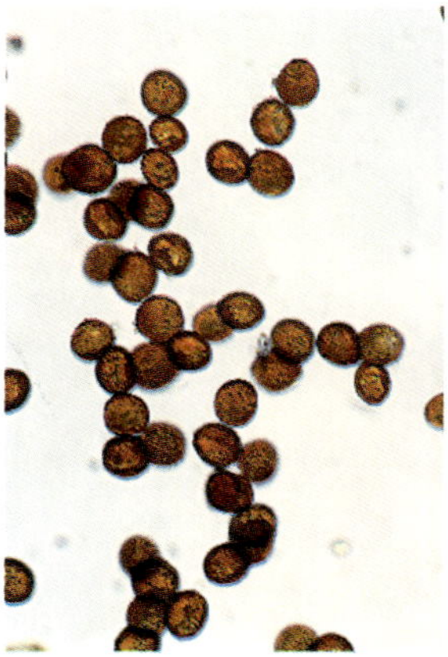

Weissia longidens **verrucose** spores.
100 µm

Philonotis scabrifolia **verticillate** shoot apex.
1 mm

Barbula unguiculata **viatical** habit.

vide — a Latin word meaning *see*.

villose — shaggy, covered with long hairs (•below) (*see* **hairy** for a comparison of similar terms).

Dawsonia superba **villose** calyptra.
10 mm

Polytrichum commune **villose** calyptra.
1 mm

Pogonatum subulatum **villose** calyptra.
1 mm

villus (plural **villi**) — in mosses, a **paraphyllium**.

vitta (plural **vittae**) (adjective **vittate**) — (1) a stripe, (2) near the center of the leaf of some leafy liverworts, one or more rows of enlarged cells running along roughly the middle of the leaf (•below, left).

voucher — a specimen preserved as a sample of material studied or illustrated for a publication.

wart — a small protuberance.

water-sac — in some leafy liverworts with **complicate-bilobed** leaves, a hollow sac-like **lobule** (•below, middle and right).

Bazzania nitida **vitta**.
100 µm

Lepidolaena palpebrifolia **water-sacs**.
1 mm

Jubulopsis novae-zelandiae **water-sacs**.
1 mm

weft — a growth form of loosely interwoven and **ascending** shoots (•below, left).

whorled — arranged in a circle or ring.

wide-spreading — making an angle of almost 90° with the stem (•below, middle) (*see* **spreading** for a comparison of similar terms).

wing — a membranous extension of a leaf, **spore**, or other structure (•below, right).

Thuidium laeviusculum **weft** growth form.
10 mm

Polytrichum juniperinum **wide-spreading** leaves.
5 mm

Mittenia plumula **wing** of leaf.
100 μm

winged thallus — a **thallus** with membranous **unistratose** extensions along one or both sides.

wiry — slender but stiff and strong.

x — the basic **haploid chromosome** number of a **species** or other **taxon**, for example n = 4. Haploid counts of 4, 5, 6, 7 (and their multiples) and also 11 are common in mosses, whereas hornworts have a count of 5 or 6, and three-quarters of liverworts have a count of 9 (*compare with* **n**, the haploid chromosome number of an organism or tissue). Bryophyte chromosomes are small and they tend to clump together, so they're often hard to count.

xeric — very dry (*compare with* **mesic** and **hydric**).

xeromorphic — structurally adapted for surviving in dry **habitats** (*compare with* **mesomorphic**, **hygromorphic**, and **hydromorphic**).

xerophilous — preferring dry **habitats** (*compare with* **mesophilous**, **hygrophilous**, and **hydrophilous**).

xerophyte — a plant adapted to dry **habitats** (*compare with* **hydrophyte**, **hygrophyte**, and **mesophyte**). Some mosses can withstand remarkably long periods of bone-dry **desiccation**—*Bryum argenteum* has been revived after nearly two years in dry storage, and *Dicranoweisia cirrata* survived 9 years and *Anoectangium compactum* 19 years.

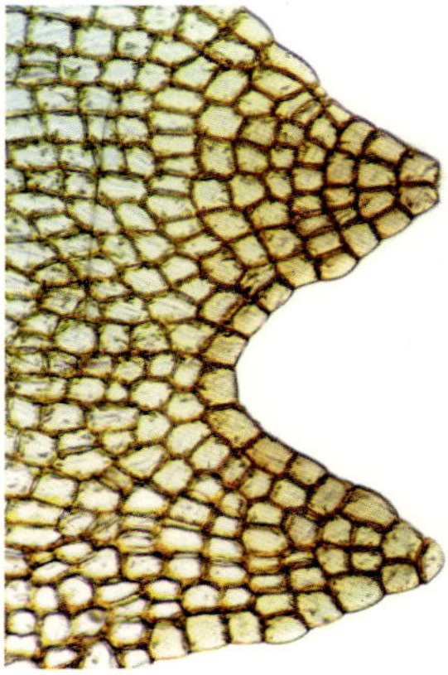

Metahygrobiella drucei **ypsiloid** sinus.
100 μm

xerophytic — said of plants or vegetation adapted to dry **habitats** (*compare with* **hydrophytic**, **hygrophytic**, and **mesophytic**).

xs — cross (**transverse**) section (*synonym* **TS**).

ypsiloid (or **ypsiliform**) — U-shaped or horseshoe-shaped (•left) (*synonym* **hyoid**) (*compare with* **digammoid** or **digammiform**, V-shaped).

***Zoopsis*-type branching** — in leafy liverworts, a type of branching in which a branch replaces an entire leaf (for a comparison with other types of branching, *see* ***Acromastigum*-type branching**).

zygomorphic — **bilaterally symmetric**, said of anything which can be cut into two mirror halves along only one plane (*the opposite of* **actinomorphic** or **radially symmetric**). Bilateral symmetry is considered more specialized than radial symmetry.

zygote — the cell (usually **diploid**) resulting from the fusion of two **gametes** (usually **haploid**).

further reading

Crum, HA (1983): *Mosses of the Great Lakes Forest* (Third Edition). University of Michigan, Ann Arbor.

Crum, HA; Anderson, LE (1981): *Mosses of Eastern North America*. Columbia University Press, New York.

Flowers, S (edited by A Holmgren) (1973): *Mosses: Utah and the West*. Brigham Young University, Provo.

Hébant, C (1977): *The Conducting Tissues of Bryophytes*. J. Cramer, Vaduz.

Inoue, H; Schuster, RM (1971): A monograph of the New Zealand and Tasmanian Plagiochilaceae. *Journal of the Hattori Botanical Laboratory* **34**, 1–225.

Niklas, KJ (1997): *The Evolutionary Biology of Plants*. University of Chicago Press, Chicago.

Noguchi, A (1987): *Illustrated Moss Flora of Japan*. Hattori Botanical Laboratory, Nichinan.

Raven, PH; Evert, RF; Eichhorn, SE (1998): *Biology of Plants* (Sixth Edition). W.H. Freeman, San Francisco.

Schofield, WB (1985): *Introduction to Bryology*. Macmillan, New York.

Schuster, RM (1966-80): *The Hepaticae and Anthocerotae of North America East of the Hundredth Meridian*, Volumes 1-4, Columbia University Press, New York & London.

Schuster, RM (1992): *The Hepaticae and Anthocerotae of North America East of the Hundredth Meridian*, Volumes 5-6, Field Museum of Natural History, Chicago.

Schuster, RM (ed) (1983-4): *New Manual of Bryology*, Volumes 1-2. Hattori Botanical Laboratory, Nichinan.

Scott, GAM; Stone, EG; Rosser, C (1976): *The Mosses of Southern Australia*. Academic Press, London.

Smith, AJE; Smith, R (1978): *The Moss Flora of Britain & Ireland*. Cambridge University Press, Cambridge.

Zander, RH; Eckel, PM (1993): *Genera of the Pottiaceae: Mosses of Harsh Environments*. *Bulletin of the Buffalo Society of Natural Sciences*, Volume 32.